T0323939

CONVENTIONAL ENERGY IN NORTH AMERICA

CONVENTIONAL ENERGY IN NORTH AMERICA

Current and Future Sources for Electricity Generation

JORGE MORALES PEDRAZA

Senior Consultant, Morales Project
Consulting, Vienna, AUT

ELSEVIER

Elsevier
Radarweg 29, PO Box 211, 1000 AE Amsterdam, Netherlands
The Boulevard, Langford Lane, Kidlington, Oxford OX5 1GB, United Kingdom
50 Hampshire Street, 5th Floor, Cambridge, MA 02139, United States

Notices
Knowledge and best practice in this field are constantly changing. As new research
and experience broaden our understanding, changes in research methods, professional
practices, or medical treatment may become necessary.

Practitioners and researchers must always rely on their own experience and knowledge
in evaluating and using any information, methods, compounds, or experiments
described herein. In using such information or methods they should be mindful of their
own safety and the safety of others, including parties for whom they have a professional
responsibility.

To the fullest extent of the law, neither the Publisher nor the authors, contributors, or
editors, assume any liability for any injury and/or damage to persons or property as a
matter of products liability, negligence or otherwise, or from any use or operation of
any methods, products, instructions, or ideas contained in the material herein.

Library of Congress Cataloging-in-Publication Data
A catalog record for this book is available from the Library of Congress

British Library Cataloguing-in-Publication Data
A catalogue record for this book is available from the British Library

ISBN: 978-0-12-814889-1

For information on all Elsevier publications visit our website at
https://www.elsevier.com/books-and-journals

Publisher: Candice G. Janco
Acquisition Editor: Marisa LaFleur
Editorial Project Manager: Karen R. Miller
Production Project Manager: James Selvam
Cover Designer: Greg Harris

Typeset by TNQ Technologies

Working together
to grow libraries in
developing countries

www.elsevier.com • www.bookaid.org

To my grandchildren Adrián and Mikail.

Contents

Preface

Writing a book on any subject is always an enormous task for the author, but when the theme is that of energy, then the challenge is even more significant.

The present book is the fifth in the series about the use of conventional, renewable, and nuclear energies for electricity generation and heating in different regions that have been prepared by me and published by several publishing houses in recent years. The books already published describe the current situation and perspectives in the use of conventional (oil, natural gas, and coal), renewable, and nuclear energies for electricity generation and heating in the Latin American and the European regions.

Following the same structure of the books already published, the present book talks about the use of conventional energies in the North American region as a whole, as well as in the US and Canada. In its different chapters, the current situation and perspectives in the use of these types of energy sources for electricity generation and heating in the entire North American region and in particular in the US and Canada are briefly described.

The present book has five chapters, 144 figures, and 34 tables. The first chapter describes the current situation and perspectives in the use of conventional energies for electricity generation and heating at the global level and at the level of the North American region considered as a whole. Based on the information included in this chapter, the electricity generation and heating in the North American region during the period 2012–17 using oil, natural gas, and coal as fuel decreased by 1,6%, falling from 2.903 billion kWh in 2012 to 2.859,2 billion kWh in 2017.

Because the US and Canada are reducing the use of coal, oil, and other liquids for electricity production, it is probable that the role of these fossil fuels as part of the energy mix of these two countries will continue to fall during the coming years. On the contrary, the use of natural gas for electricity generation and heating is expected to increase in the energy mix of these two countries during the coming years.

The second chapter describes the current situation and perspectives in the use of oil for power generation in the North America region as a whole, as well as the role that this type of conventional energy plays and will continue to play in electricity generation and heating in the US and Canada. According to the information included in this chapter, it is projected that up to 2040 the share of oil and other liquids in the world electricity generation will decrease by 45,5%, falling from 1,1 TWh in 2012 to 0,6 TWh in 2040. The percentage of participation of oil and other liquid fuels in the world's electricity generation and heating is projected to decrease from 5% in 2012 to less than 1,6% in 2040; that is a reduction of 3,4%. It is expected that this trend will continue after 2040. In the specific case of the North America region, the participation of oil and other liquids in the energy mix of the US and Canada is today the lowest considering the three types of fossil fuels used for electricity generation and heating, and it is foreseen that this situation will not change during the coming years.

In the specific case of the US, the tendency is to reduce the use of petroleum coke for electricity generation and heating during the coming years, as well as its role in the energy mix of the country. Within the period 2010—17, the peak in the consumption of petroleum coke for electricity generation and heating in the US was registered in 2011 and since that year the use of petroleum coke has been falling every year. The leading consumer of this type of energy source in the US is the electric power sector itself. In 2017, this sector consumed 81,5% of the total electricity generated in the country. The commercial sector is the area with the lowest consumption of petroleum coke for electricity generation and heating in the US during the whole period under consideration. In 2017, this sector consumed only 0,09% of the total.

It is important to highlight that due to the negative impact on the environment and the population as a result of the use of oil for electricity generation and heating, during the period 2010—16 the number of oil-fired power plants used in the US for this purpose dropped significantly (8%), falling down from 1.169 power plants in 2010 to 1.076 in 2016. It is forecast that more oil-fired power plants will be closed in the US during the coming years.

In the case of Canada, the electricity generation using oil as fuel decreased by 38% during the period 2010—17, falling from 6,9 TWh in 2010 to 4,3 TWh in 2017. It is predictable that the electricity generation in the country will continue following this trend during the coming years but at a slow rate. Different energy measures adopted by the Canadian government to reduce the use of oil for electricity generation and heating and the negative impact that the use of this type of energy source has on the environment and the population will be the leading causes for this situation.

Chapter 3 describes the current situation and perspectives in the use of natural gas for global electricity generation and heating, highlighting not only the case in the North American region as a whole but also in the US and Canada. Natural gas is the second most commonly used fossil fuel in the world for power generation and the only one whose share of primary energy consumption is projected to grow during the coming years in many countries, including the US and Canada. Without a doubt, the natural gas market in the North America region is a mature market, with a financial structure well integrated and with a high national production in Canada and the US. It is expected that this trend will continue during the coming years.

In the US, the net electricity generation using natural gas and other gases as fuel in 2017 in all sectors of its economy reached the amount of 1.287.053 thousand MWh. Per sectors, the electricity generation using natural gas and other gases as fuel in the US in 2017 included:

- Electricity utilities: the electricity generation from natural gas was 617.725 thousand MWh and from other gases was 164 MWh.
- Independent power producers: the electricity generation from natural gas was 558.439 thousand MWh and from other gases was 4.013 thousand MWh.
- Commercial sector: the electricity generation from natural gas was 7.516 thousand MWh.
- Industrial sector: the electricity generation from natural gas was 89.188 thousand MWh and from other gases was 9.982 thousand MWh.

The number of gas-fired power plants used for electricity generation and heating in the US increased by 8,3% during the period 2006—16, rising from 1.705 power plants in 2006 to 1.846 power plants in 2016. As a result of the measures adopted by the US administration in the energy sector, it is anticipated that the number of natural gas-fired power plants in the US will continue to grow during the coming years.

In the case of Canada, the electricity generation using natural gas as fuel during the period 2010—17 increased by 53,6%, rising from 47,8 TWh in 2010 to 73,4 TWh in 2017. The peak in the electricity generation and heating using natural gas as fuel during the period considered was reached in 2017. In 2017, the total electricity generated in Canada using all available energy sources reached the total of 693,4 TWh. In that year, the power produced using natural gas as fuel represented 10,6% of the total. It is expected that the use of natural gas for electricity generation and heating in the country will continue to grow during the coming years, as a result of the closure of several old and inefficient coal-fired power plants; it is foreseen that the role of natural gas in the energy mix of the country will be higher than today. It is also forecast that in 2035 the total electricity generated by gas-fired power plants in Canada will reach 114.473 GWh; this represents an increase of 80,8% with respect to 2016.

Chapter 4 describes the current situation and perspectives in the use of coal for electricity generation and heating not only at the level of the North American region but also in the US and Canada. Most of the coal from the North American region was consumed in the US in 2017, which accounted for 94,7% of the region's total coal use in 2017 (350,7 million tons equivalent). It is expected that, as a result of the measures adopted by the US administration, the use of coal for electricity generation and heating will remain relatively flat until 2040 rising by only 2 quadrillions Btu during the whole period. However, if the proposed CPP were implemented, then the US coal consumption is expected to decline by almost 3 quadrillions Btu by 2040.

Moreover, strong growth in shale gas production, decreasing electricity demand as a result of the implementation of different energy efficiency measures, the adoption of environmental regulations to reduce the negative impact of the use of some fossil fuel for electricity generation and heating, and the increased use of renewable energy sources for electricity generation and heating are expected to reduce the share of coal-fired power generation within the total US electricity production (including electricity generated at plants in the industrial and commercial sectors) from 37% in 2012 to 26% in 2040; this means a decrease of 9% for the whole period. Finally, it is important to stress that the number of coal-fired power plants used for electricity generation and heating in the US decreased by 56,8% during the period 2010—17, falling from 580 coal-fired power plants in 2010 to 251 coal-fired power plants in 2017. It is expected that this trend will continue at least during the coming years but at a slower pace. Old and inefficient coal-fired power plants will be closed.

In the case of Canada, coal is mainly used for electricity generation and heating which accounts for about 85,3% of the total coal consumption; 7,2% goes to coke manufacturing and various industries; and 7,6% to other non-energy uses. Coal-fired generation contributes about 10% of Canada's total electricity generation. Without a doubt, coal is expected to play a relatively minor role in Canada's energy supply system in the future, due to the adoption of strong measures to protect the environment and the population. For this reason, the position of coal within the energy supply system in Canada is expected to decline further, due to the adoption of federal and provincial government measures to reduce emissions of several contaminating gases produced as a result of burning coal for electricity generation and heating.

Canada's total coal consumption is expected to decline by 51% (0,4 quadrillions Btu) during the period 2012—40, and the share of coal in the total primary energy supply is likely to decrease from 5% in 2012 to 2% in 2040; this means a reduction of 3% for the whole period. It is important to highlight that in 2017, coal was the third energy source for power generation in Canada with 10,6% of the total electricity produced in the country in that year (693,4 TWh, including the electricity generated by the industry), after hydro with 60,7% and nuclear power with 16,8%. However, the participation of coal in the energy mix of the country will continue to drop during the coming years as a result of the closure of old and inefficient coal-fired power plants.

Chapter 5 summarizes the essential aspects that describe the current situation and perspectives in the use of oil, natural gas, and coal for electricity generation and heating throughout the North American region as a whole, and especially in the US and Canada.

Acknowledgments

During the preparation of the present book Mr. Alejandro Seijas Lopez, Master of Engineering in Energy and Mechanical Engineering Degree, gave me an essential assistance in the compilation of relevant information regarding the current and future role of fossil fuels in the electricity generation in the North America region, and in the preparation of two chapters of the book.

Without any doubt, the present book is a reality thanks to the valuable support of my lovely wife, Aurora Tamara Meoqui Puig, who had assumed other family responsibilities to give me the necessary time and the adequate environment to write the present book.

Introduction

Electricity consumption is an essential component of the modern life. It not only provides clean and safe light throughout the day, but also in many countries refreshes homes on hot summer days, and in others warms them in winter. In all countries, it allows the use of electrical and electronic equipment in which the use of electricity is essential to ensure their proper functioning. Although hundreds of millions of Americans and Canadians connect to the power grid every day, most of them do not think about how they get the electricity consumed, and how much it costs to produce it. Keeping the North America region energized is actually an amazing feat, a daily miracle.

CHAPTER 1

General Overview of the Energy Sector in the North America Region

Contents

Conventional Energy in North America
ISBN 978-0-12-814889-1
https://doi.org/10.1016/B978-0-12-814889-1.00001-2

1

Introduction

Electricity generation and consumption are essential components of modern life. They not only provide clean and safe light throughout the day, but also in many countries, especially those located in the hot areas on the planet, refresh homes on summer days, and, in countries located in the coldest regions, warm them in the winter. In all states, independent of their locations and economic development, they allow the use of electrical and electronic equipment in which the use of electricity is essential to ensure their proper functioning. Without electricity, it is impossible to use mobile phones, tablets, and computer equipment. Although hundreds of millions of Americans and Canadians connect to the power grid every day, most of them do not think or have an idea about how they get the electricity consumed, and how much it costs to produce it. When people stop to analyze this issue, it is not difficult to conclude that keeping the North America region energized is a fantastic achievement, a daily miracle (History of Electricity, 2014).

It is a reality that without energy, people will be deprived of heating, cooling, and light in their homes and workplaces. They would not have access to television and the internet, among other services relevant to their lives and professional activities. With the development of modern societies, the use of these devices and services is expected to grow, as well as the dependence on them. This situation is not only the characteristic of developed countries, but also of the developing countries as well, particularly for the most advanced developing countries such as China, South Africa, India, South Korea, Brazil, and Russia, to mention some examples. The dependence of these countries on the use of electricity grows every year, and it is expected that this trend will continue without change during the coming decades.

The growth in the world economy means that more energy will be required,[1] particularly power in the form of electricity to guarantee the functioning of the different industries, hospitals, cinemas, theaters, universities, and service centers, to mention a few examples. Even though the consumption of nonfossil fuels is expected to grow faster than the use of fossil fuels, this last type of energy source is still expected to account for 78%

---

[1] According to the BP Energy Outlook 2016 report, the current world energy growth is 1,4% per annum, which is 0,9% lower than the growth registered during the period 2000—14 (2,3%).

of energy consumption in 2040, according to the International Energy Outlook 2016 with Projection to 2040 report.

Fossil fuels[2]—coal, petroleum (oil), and natural gas—are concentrated organic compounds found in the Earth's crust, and are formed from the remains of plants and animals that lived millions of years ago in the form of concentrated biomass. For many purposes, oil and natural gas resources are classified into four categories. According to the Technically Recoverable Shale Oil and Shale Gas Resources (2015) paper, the categories are:

- remaining oil and gas-in-place: original oil and gas in-place minus cumulative production at a specific date;
- technically recoverable resources: includes all the oil and gas that can be produced using the current technology, industry practice, and geologic knowledge;
- economically recoverable resources: the portion of technically recoverable resources that can be commercially produced; and
- proved reserves: volumes of oil and gas that geologic and engineering data demonstrate with reasonable certainty to be recoverable in future years from known reservoirs under current economic and operating conditions (Fig. 1.1).

The crude oil and natural gas volumes reported for each type of energy resource category are the best estimates that can be elaborated based on a combination of facts and assumptions regarding:

- the geophysical characteristics of the rocks;
- the fluids trapped within those rocks;
- the capability of the extraction technologies used;
- the prices received; and
- the costs paid to produce oil and natural gas.

However, not all fossil fuel estimates are based on the same combination of facts and assumptions mentioned above. For example, oil and natural gas-in-place estimates are based on fewer facts and more assumptions,

---

[2] The consumption of different fossil fuels, specifically natural gas and coal for electricity generation and heating, and oil for the same purpose but mainly for the movement of different transport systems, makes possible the continuous development of modern societies. Through the consumption of this type of energy sources different transport systems are operated, as well as the functioning of equipment for the production of steam and electricity generation, among others. Through the consumption of this type of energy source, it is possible also to produce commercial, industrial, medical, and many other different types of products, indispensable to guarantee the welfare of the people.

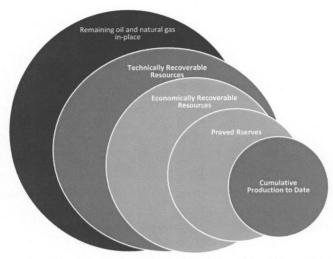

Figure 1.1 Stylized representation of oil and natural gas resource categorizations. *(Source: Energy Information Administration (EIA).)*

while proved reserves are based mostly on facts and fewer assumptions (Today in Energy, 2018).

It is important to highlight that according to the Petroleum Resources Management System (2007) report, "the estimation of petroleum resource quantities involves the interpretation of volumes and values that have an inherent degree of uncertainty." The same thing can be said in the case of natural gas. For this reason, it is impossible to provide, with the current technology, precise volumes and values of either oil or natural gas at any given place and time.

Within all available fossil fuels, natural gas is expected to be the fastest-growing fossil fuel in the projection period until 2040. Global natural gas consumption is projected to increase by 1,9% per year until 2040. Abundant natural gas resources and robust production, including rising supplies of tight gas, shale gas, and coalbed methane, are expected to contribute to the strong competitive position of natural gas in comparison to oil and coal within the energy mix of several countries. Although liquid fuels—mostly petroleum-based—are expected to remain the largest source of world energy consumption, their share of world marketed energy consumption is projected to drop by 3% during the period 2012—40 falling from 33% in 2012 to 30% in 2040 (International Energy Outlook 2016 with Projection to 2040, 2016).

Several factors will contribute to the decline in the consumption of liquid fuels within the energy mix of many countries. One of these factors is the foreseeable rise of oil prices in the long-term, which will lead many energy users to adopt more energy-efficient technologies and to switch away from liquid fuels to the consumption of another energy source for electricity generation and heating, when feasible. Coal is expected to be the world's slowest-growing energy source during the projected period. It is expected to rise by 0,6% per year and is likely to be surpassed by natural gas by 2030 (International Energy Outlook 2016 with Projection to 2040, 2016).

According to the report mentioned above, world oil consumption and other liquid fuels are expected to grow from 90 million barrels per day in 2012 to 100 million barrels per day in 2020, and 121 million barrels per day in 2040; this means an increase of 34,4% for the whole period considered. However, it is important to single out that most of the growth in liquid fuels consumption is not in the electricity generation sector but the transportation and industrial sectors. In the transportation sector, liquid fuels will continue to provide most of the energy consumed at least during the foreseeable future.

The high growth registered in the past years in the use of different renewable energy sources for electricity generation and heating in several countries, the rise in energy investment in many of them, and the increase in new energy capacities at world level have changed the landscape for the energy sector worldwide. There has been not only an accelerated growth in the use of different renewable energy sources for electricity generation and heating in many countries but also an improvement in the evolution of the technology associated with this type of energy sources. This development has contributed not only to the fall in prices of many renewable energy sources used for electricity generation and heating, but also to a greater separation between the economic growth of the different countries, the greenhouse gas emissions produced using specific energy sources, and the abrupt climatic changes that are taking place worldwide. "Most countries have achieved a more diversified energy mix with growth in community ownership and evolution of microgrids" (see Figs. 1.2 and 1.3) (World Energy Resources 2016, used with the permission of the World Energy Council, www.worldenergy.org).

According to Figs. 1.2 and 1.3, the participation of oil in the world primary energy consumption has been increased during the period under consideration by 0,71%, rising from 33,49% in 2010 to 34,2% in 2017. It is expected that this trend will continue without change during the

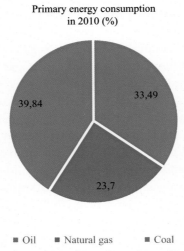

Figure 1.2 Primary energy consumption in 2010. *(Source: BP Statistical Review of World Energy 2016, 2016. 65th ed.)*

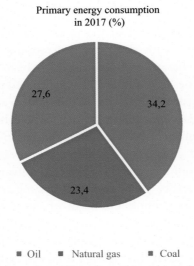

Figure 1.3 Primary energy consumption in 2017. *(Source: BP Statistical Review of World Energy 2018, June 2018.)*

coming years. On the other hand, the participation of natural gas in the world primary energy consumption has been decreased by 0,3% falling from 23,7% in 2010 to 23,4% in 2017. However, it is projected that the participation of natural gas in the world primary energy consumption will increase during the coming years. Finally, the involvement of coal in the world primary energy consumption has been reduced by 12,24% falling from 39,84% in 2010 to 27,6% in 2017. It is predictable that the participation of coal in the world primary energy consumption will continue to fall in the coming years, as a result of the implementation of more strict measures to reduce the negative impact on the environment and the population for the use of coal for electricity generation and heating.

The International Energy Outlook 2016 report suggests that world crude oil production should increase by 30,5 million barrels per day, to meet the future world crude oil demand. For this reason, crude oil production should register a steady growth in the Organization for Economic Co-operation and Development (OECD) (4,2% or 1,1 million barrels per day), in the Organization of the Petroleum Exporting Countries (OPEC) (3,7% or 1,3 million barrels per day), and an average lower growth in other producing countries (an increase of 1,3% or 0,4 million barrels per day). According to EIA estimates, petroleum and other liquid fuels production in countries outside of the OPEC grew by 1,4 million barrels per day in 2015; the main growth occurred in the North America region.

The most substantial increases in natural gas production during the period 2012−40 are expected to occur in non-OECD Asia (18,7 trillion cubic feet), the Middle East (16,6 trillion cubic feet), and the OECD Americas (15,5 trillion cubic feet). In the US, the production of natural gas is expected to increase by 11,3 trillion cubic feet and will come mainly from shale resources (more than half of the US natural gas production), according to the International Energy Outlook 2016 with Projection to 2040 report. Without a doubt, "shale gas and tight oil are revolutionizing world energy markets. New drilling methods and technologies have suddenly given North America access to vast deposits of oil and shale and tight rock formations. These resources, largely inaccessible only a decade ago, represent a significant source of economic growth, jobs, and tax revenue" (Dempster et al., 2016).

Also, it is expected that in the US the production of tight gas, shale gas, and coalbed methane will substantially increase until 2040. "The application of horizontal drilling and hydraulic fracturing technologies has made it possible to develop the US shale gas resource, contributing to a near doubling of estimates for total US technically recoverable natural gas resources over the past decade" (International Energy Outlook 2016 with Projection to 2040, 2016).

Tight gas, shale gas, and coalbed methane resources are expected to account for about 80% of the total production in 2040 in Canada and China.

On the other hand, liquefied natural gas (LNG) is expected to account for a growing share of world natural gas trade rising from about 12 trillion cubic feet in 2012 to 29 trillion cubic feet in 2040; this represents an increase of 17 trillion cubic feet for the whole period considered or 141,7%. Most of the rise in liquefaction capacity is foreseeable to occur in Australia and in the North America region, where a multitude of new liquefaction projects are planned or under construction (International Energy Outlook 2016 with Projection to 2040, 2016).

According to the report mentioned above, natural gas production in the OECD Americas is expected to grow by 49% during the period 2012—40. The US, which is the largest natural gas producer in the OECD Americas and also in the OECD as a whole, accounts for more than 45,8% of the expected region's total production growth during the period under consideration. That grow is believed to raise the production of natural gas from 24 trillion cubic feet in 2012 to 35 trillion cubic feet in 2040; this represents an increase of 11 trillion cubic feet or 45,8%.[3] In 2040, shale gas is expected to account for 55% of the total US natural gas production, tight gas is expected to account for 20%, and offshore gas production from the lower 48 states is likely to account for 8%. The remaining 17% is likely to come from coalbed methane, Alaska, and other associated and nonassociated gas coastal resources in the lower 48 states.

Coal is the most abundant of all available fossil fuels in the world and is used by a variety of economic sectors, including the electricity generation sector,[4] for iron and steel production, for cement manufacturing, and as a liquid fuel. Compared with the strong growth in coal consumption in the early 2000s, the worldwide coal consumption is projected to remain flat between 2015 and 2040 (about 160 quadrillion Btu). Coal-fired power plants currently produce 40% of the world's electricity and are forecast to continue to participate, in a significant manner, in the energy mix of several

---

[3] US shale gas production is expected to grow 100% during the period 2012—40, rising from 10 trillion cubic feet in 2012 to 20 trillion cubic feet in 2040, offsetting declines in production of natural gas from other US sources (International Energy Outlook 2016 with Projection to 2040, 2016).

[4] In the specific case of electricity generation and heating, the current level of coal participation will almost be the same at least during the coming years in several countries but in others its role will be much less than today. Also, it is important to clarify that not all types of coal are used for electricity generation and heating. Steam coal or lignite is used mainly for power generation and heating while coking coal is mainly used for iron and steel production.

countries over the next three decades. However, it is expected that the role of coal in the energy mix of many countries, including the US and Canada, will be lower than the level reached today.

At this moment, the largest coal producing countries are China, the US, India, Indonesia, Australia, and South Africa. India's coal consumption is expected to continue to grow by an average 2,6% per year during the period 2015—40, surpassing the US as the second-largest coal consumer before 2020 (International Energy Outlook 2017, 2017).

It is essential to be aware that coal is known as the most carbon-intensive fossil fuel and the continuing use of coal in global electrification could have negative implications on the environment and human health (World Energy Resources 2016, used with the permission of the World Energy Council, www.worldenergy.org). For this reason, the use of coal for electricity generation and heating is increasingly replaced by natural gas, several types of renewable energy sources, and nuclear power by some countries. The world industrial demand for coal is also decreasing, and this trend is expected to continue during the coming years.

According to the BP Energy Outlook 2016 Outlook to 2035 (2016) report, fossil fuels currently meet 81% of the US energy demand, and they are expected to remain the dominant source of energy powering the global economy,[5] providing around 60% of the growth in energy source and accounting for almost 80% of the total energy supply in 2035, down from 86% in 2014 (a reduction of 6%). In 2040, fossil fuels will still account for 78% of energy use.

---

[5] Oil has remained, since it was discovered in the US in the 18th century, as the world's main energy source consumed. Despite all development that has occurred in the different energy sources over the past decades, the substitution of oil with other less polluting energy sources is not yet imminent. The use of natural gas is the second source of energy used in the generation of electricity worldwide. It is important to note that natural gas is projected as the only fossil fuel whose participation in primary energy consumption is expected to grow in the coming years at world, regional, and local levels. Coal represents the third fossil fuel for electricity generation worldwide. However, this role at the world level will be lower in the coming years as a result of the actions recently approved by the international community to reduce the negative impact of the consumption of fossil fuels, mainly coal, for electricity generation, the need to mitigate climatic changes that is increasingly affecting several countries, and the need for a more accelerated transition to the use of other forms of cleaner energy in multiple economic activities, among others (World Energy Resources 2016 used with the permission of the World Energy Council, www.worldenergy.org).

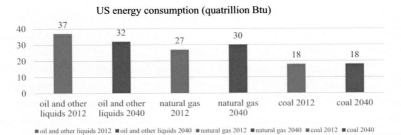

Figure 1.4 Participation of various energy sources in the US energy mix. *(Source: EIA database 2018.)*

The future participation of the different fossil fuels in the US energy mix is shown in Fig. 1.4.

According to Fig. 1.4, the consumption of oil and other liquids in the US is expected to have the higher percentage in 2040 (32%) followed closely by natural gas (30%) and coal (18%). Although liquid fuels, mostly being oil-based now, are expected to remain the largest source of world energy consumption in the coming years, the liquids share energy consumption in the US is expected to fall by 5% from 37% in 2012 to 32% in 2040. In the specific case of natural gas, it is likely that this type of energy source increases its participation in the country energy mix during the period 2012–40 rising from 27% in 2012 to 30% in 2040; this means an increase of 3%. Finally, it is expected that the participation of coal in the US energy mix will not change during the whole period considered.

In the case of Canada, it is important to single out that abundant energy resources located within the country have contributed to Canada's position as one of the world's largest energy producers.[6] However, the "long-term success of shale oil and gas greatly depends on the industry's response to the general public's environmental concerns" (Dempster et al., 2016).

According to the Canada's Energy Future 2017 report, by 2040 it is expected that the Canadian crude oil production will reach 6,3 million barrels per day. That is 59% higher than the crude oil production achieved in 2016 (4 million barrels per day).

---

[6] "Economic, social, and environmental issues are limiting the potential of Canada's shale oil and gas industry. Whereas British Columbia and Alberta have promoted investments in shale energy operations, other provinces have taken a step back to further evaluate their development approaches" (Dempster et al., 2016).

On the other hand, natural gas production in Canada is expected to grow by 1,2% per year on an average over the period 2012—40, raising from 6,1 trillion cubic feet in 2012 to 8,6 trillion cubic feet in 2040. In Canada, like in the US, much of the production growth is expected to come from growing volumes of tight gas and shale gas production (International Energy Outlook 2016 with Projection to 2040, 2016).

The abundance of shale oil and gas in Canada "is helping to push down petroleum prices, adding further criticism that higher production is discouraging people from transitioning toward environmentally friendly fuel sources, such as solar, geothermal, and biofuel.[7] On the other hand, low natural gas prices are encouraging more people to convert to clean-burning natural gas" (Dempster et al., 2016).

Finally, in the case of thermal and metallurgical coal,[8] the following can be stated according to the Canada's Energy Future 2017 report: the total Canadian coal production in 2015 was 61,9 million tons. As the fuel for coal-fired power plants, in 2015, thermal coal production accounted for 93% of the total coal consumption in Canada. However, during the period 2016—40, the demand for thermal coal is expected to decline to 86%, falling from 37,7 million tons in 2016 to 5,4 million tons in 2040.[9] This declining trend is driven primarily by retirements of the coal-fired power plants capacity resulting from regulations to phase out old and inefficient coal-fired power plants by 2030. In response to declining domestic coal demand, production of thermal coal in Canada during the period 2016—40 is also expected to drop 78,2% falling from 36,3 million tons in 2016 to 7,9 million tons in 2040.

---

[7] The transition to a low-carbon energy society will require the use of more renewable energy sources than previously thought, if current levels of energy consumption per capita and lifestyles are to be maintained.

[8] In Canada, "metallurgical coal is primarily used for steel manufacturing. Much of Canada's metallurgical coal production is exported and future production trends are linked to global coal demand and prices." However, domestic demand for metallurgical coal used in steel manufacturing is expected to be stable over the period 2016—40 at 3 million tons. Global demand for metallurgical coal is expected to grow moderately over the period 2016—40, resulting in a steady growth in net exports from Canada. The total metallurgical coal production in Canada is expected to increase 13,4% from 26,9 million tons in 2016 to 30,5 million tons in 2040 (Canada's Energy Future 2017, 2017).

[9] Most of the thermal coal consumed in Canada in 2040 for electricity generation and heating will be in coal-fired power plants but equipped with CCS technology.

The World Crude Oil Sector

Crude oil, also called "petroleum", is used for a wide range of applications, including for electricity generation and heating, in the transport sector, for the production of plastic and synthetics and other products, and for powering internal combustion engines, among others. Bitumen, the thickest form of petroleum, is used for paving roads, forming the blacktop and roofing. Almost all countries in the world use crude oil for some of the above-mentioned activities.

The evolution of the production of crude oil at world level during the period 2010—17 is shown in Fig. 1.5.

According to Fig. 1.5, the production of crude oil at world level increased by 11,2%, rising from 83.325 thousand barrels per day in 2010 to 92.649 thousand barrels per day in 2017. It is important to highlight that the world production of crude oil increased each one of the years within the period considered. It is expected that the production of crude oil at world level will continue to grow at least during the coming years.

On the other hand, the consumption of crude oil at world level has been increasing during the whole period 2010—17, and this trend is expected to continue during the coming years. The evolution of the consumption of crude oil at world level during the period 2010—17 is shown in Fig. 1.6.

By the data included in Fig. 1.6, the following can be stated: the consumption of crude oil at world level increased by 10,9% during the period considered, rising from 88.535 thousand barrels per day in 2010 to 98.186 thousand barrels per day in 2017. Looking at Fig. 1.6, it is easier to see that the consumption of crude oil at world level increased each one of the years included in the period considered. Based on this fact, it can be expected that the current trend in the use of crude oil will continue to

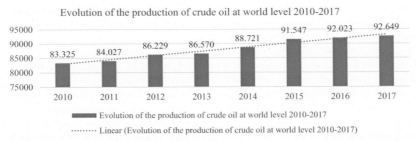

Figure 1.5 Evolution of the production of crude oil at world level during the period 2010—17. *(Source: BP Statistical of World Energy 2018, June 2018.)*

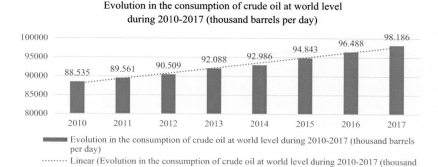

Figure 1.6 Evolution of the consumption of crude oil at world level during the period 2010–17. *(Source: BP Statistical Review of World Energy 2018, June 2018.)*

increase at least during the coming years. However, it is likely that the use of crude oil for electricity generation and heating will be lower than today.

Several factors could reduce the crude oil demand growth rate in the coming years in many countries. Among these factors, the following three are considered relevant:

- decline in population in several countries, particularly within the OECD countries;
- adoption of new energy efficiency measures by governments and the industrial sectors that use crude oil in their productions; and
- increase in dependency on the use of renewable energy sources for electricity generation and heating.

It is important to note that approximately 63% of today's crude oil consumption comes from the transport sector, and this percentage is expected to increase during the coming years. At the same time, the use of crude oil for electricity generation and heating is expected to continue to fall in many countries until 2040 and in many others beyond that year.

According to the World Energy Resources 2016 report, "oil remained the world's leading fuel, accounting for 32,9% of the global energy consumption". However, it is expected that crude oil participation in the world energy mix will be lower in the future in comparison with the level reported today, particularly for electricity generation and heating. By region, crude oil demand in OECD North America, in 2016, was the highest among all regions with 24,3% of the world crude oil demand, followed very closely by Asia with 23,8%. Nevertheless, it is predictable that this situation will change in 2019. In that year, Asia will have the highest crude oil demand with 25,2% of the world crude oil demand, an increase of 1,4% concerning 2016, followed by the North America region with 24,5%, an increase of only 0,2% with respect to the level registered in 2016. Emerging economies now account for 58,1% of the global energy consumption and the global demand for liquid hydrocarbons is expected to continue to grow during the coming years (BP Statistical Review of World Energy 2016, 2016).

The evolution of the world proven crude oil reserves by country and region during the period 2010—16 is shown in Table 1.1.

According to Table 1.1, the world proved crude oil reserves increased 2,3% during the period 2010—16. The North America region increased its proved crude oil reserves by 31,9% during the period considered. In the specific case of the US, its proved crude oil reserves during the period 2010—16 increased 38,9%, while the proved crude oil reserves in Canada decreased by 7,2%. The world current and expected crude oil demand by regions during the period 2014—18 is shown in Table 1.2.

From Table 1.2, the following can be stated: the growth in crude oil demand within the North America region is expected to be 1,6%, increasing from 24,1 million barrels per day in 2014 to 24,5 million barrels per day in 2018. The other significant growth is supposed to be registered by "other Asia" with 13,2%, and China with 10,5%. OECD Europe and OECD Asia-Ocean are the only two regions that are expected to report a decrease in crude oil demand during the period considered.

On the other hand, it is important to highlight that, despite all efforts made to date by the oil industry, the substitution of crude oil to meet the energy needs of different sectors of the economy is not going to happen anytime soon "and is not expected to reach more than 5% over the next five years. Unconventional oil recovery accounts for 30% of the global recoverable crude oil reserves and oil shale resources contain at least three times as much oil as conventional crude oil reserves, which are projected at around 1,2 trillion barrels" (World Energy Resources 2016, used with the permission of the World Energy Council, www.worldenergy.org, 2016).

It is also important to single out that more than half of the increase in global energy consumption is used for power generation due to the measures adopted toward global electrification. For this reason, the share of

Table 1.1 Evolution of the World Proven Crude Oil Reserves by Country and Regions During the Period 2010–16 (million barrels).

Country/Region	2010	2011	2012	2013	2014	2015	2016
North America	27.469	30.625	34.661	37.652	40.503	36.281	36.218
United States	23.267	26.544	30.529	33.371	36.385	32.318	32.318
Canada	4.202	4.081	4.132	4.281	4.118	3.900	3.900
Latin America	334.008	336.996	338.356	341.522	341.296	342.549	339.645
Eastern Europe and Eurasia	117.310	117.314	119.881	119.874	119.863	119.860	119.856
Western Europe	12.576	10.880	10.800	11.337	10.761	10.064	11.353
Middle East	794.595	797.155	799.132	802.958	802.512	802.848	807.730
Africa	125.623	125.441	128.291	128.070	127.254	127.969	128.359
Asia and the Pacific	47.227	47.322	47.552	47.860	48.197	48.385	49.003
Total	1.458.808	1.465.733	1.478.673	1.489.272	1.490.386	1.487.893	1.492.164

Source: OPEC database 2017.

Table 1.2 World Current and Expected Crude Oil Demand by Regions During the Period 2014–18 (million barrels per day).

Region	2014	2015	2016	2017	2018
OECD Americas	24,1	24,2	24,3	24,4	24,5
OECD Asia-Ocean	8,1	8,0	7,9	7,9	7,9
OECD Europe	13,4	13,3	13,3	13,2	13,1
Another Europe	0,7	0,7	0,7	0,7	0,7
China	10,4	10,6	10,9	11,2	11,5
Other Asia	12,1	12,5	12,9	13,3	13,7
Latin America	6,8	6,9	7,0	7,1	7,2
Middle East	8,1	8,3	8,5	8,8	9,0
Africa	3,9	4,1	4,2	4,4	4,5
Others	4,8	4,6	4,7	4,7	4,8
World	92,4	93,3	94,5	95,7	96,9

Source: World Energy Resources 2016, used with the permission of the World Energy Council, www. worldenergy.org.

energy used for power generation will rise from 42% today to 45% by 2035[10]; this represents an increase of 3% for the whole period under consideration. In the specific case of crude oil, the growth in the global consumption of liquid fuels is driven by the transport sector (63% today to 88% in 2035) and by the industry sector, especially in petrochemicals, with an increase of 1,2 million barrels per day in 2035. The transport sector will account for almost two-thirds of the rise (BP Energy Outlook 2016, 2016).

The World Natural Gas Sector

Natural gas is the second most-used energy source for electricity generation and heating at world level, and it is the only fossil fuel whose participation in primary energy consumption is expected to grow during the coming years. For this reason, the use of natural gas for this specific purpose is expected to play a relevant role in the global energy transition toward a cleaner, cheaper, and safer energy during the coming decades.

Advances in technologies associated with the production and consumption of natural gas have created new prospects for more affordable and

---

[10] It is important to single out that electricity generation is the main sector where all types of energy sources compete. Without a doubt, renewables and natural gas are the two types of energy sources with the major impact in the structure of the energy mix of many countries, surpassing coal in the majority of them. This situation will not change at least until 2040.

safer natural gas supplies. The markets of this specific energy sources are increasingly interconnected as a result of gas-to-gas pricing, short-term trade, and the bargaining power of consumers.

However, the future of the world natural gas demand is highly uncertain. For this reason, new policy frameworks and continued cost reduction will be needed to make natural gas more competitive concerning other energy sources used for electricity generation and heating as well. The construction of infrastructure, the support of governments, and the closing of regulatory gaps, among others, are necessary elements to unlock the socioeconomic and environmental benefits related to the production and consumption of natural gas.

Relevant data on natural gas reserves and production at world level and by region are included in Table 1.3.

According to Table 1.3, the North America region has the lowest R/P ratio (13 years) in comparison to other areas. This weak relationship has a negative impact on the level of production of natural gas in the region. The Middle East is the region with the maximum R/P ratio with 129,5 years, followed by Africa with 66,4 years, and Europe and Eurasia with 57,4 years.

The world consumption of natural gas is expected to increase during the period 2012—40 rising from 120 trillion cubic feet in 2012 to 203 trillion cubic feet in 2040; this means an increase of 69,2%. By type of energy source, natural gas is expected to register the most significant increase in the world's primary energy consumption at least until 2040. The abundant resources of natural gas worldwide and the efficiency in the electricity production using this type of energy source as a fuel in comparison to oil

Table 1.3 Natural Gas Reserves and Production at a World Level and by Region in 2015.

Region	Proved Reserves (billion cubic feet) 2015	Production (billion cubic feet) 2015	R/P Ratio (years)
Africa	496.666,5	7.479,2	66,4
Asia Pacific	552.607,7	19.658,2	28,1
Europe and Eurasia	2.005.109,3	34.955,2	57,4
Latin America and the Caribbean	268.091	6.302,1	42,5
Middle East	2.826.617,7	21.821,1	129,5
North America	450.326	34.750,4	13
Total	6.599.418,2	124.966,2	52,8

Source: World Energy Resources 2016 used with the permission of the World Energy Council, www. worldenergy.org.

and coal contribute to the strong competitive position of natural gas among other forms of energy resources used for the same purpose (International Energy Outlook 2016, 2016).

In several countries around the world, natural gas continues to be an essential fossil fuel used not only for electricity generation and heating but in the output in the industrial sector as well. In the electricity generation sector, natural gas is an attractive option for the construction of new natural gas-fired power plants, given its moderate capital investment cost and attractive prices. Also, natural gas-fired plants operate with a relatively high level of efficiency in comparison with other types of power plants, and natural gas is the less contaminated fossil fuel that can be used for electricity generation and heating. For this reason, natural gas can displace without any difficulty the consumption of coal and other liquid fuels for this specific purpose easily and efficiently (International Energy Outlook 2016 with Projection to 2040, 2016).

Global natural gas producers are expected to increase the supply of this type of energy source by 69,2% to meet the growing demand for natural gas projected until 2040. It is also likely that the most substantial increases in natural gas production will occur in Asia, especially in those countries outside the OECD (18,7 trillion cubic feet), followed by the countries in the Middle East (16,6 trillion cubic feet), and in the OECD America (15,5 trillion cubic feet including Mexico and 13,8 trillion cubic feet without Mexico) (International Energy Outlook 2016 with Projection to 2040, 2016).

The evolution of the production of natural gas at world level during the period 2010—17 is shown in Fig. 1.7.

According to Fig. 1.7, the production of natural gas at world level during the period 2010—17 increased by 16,1%, rising from 111,92 billion cubic feet in 2010 to 129,97 billion cubic feet in 2017. It is important to

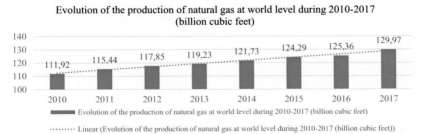

Figure 1.7 Evolution of the production of natural gas at world level during the period 2010—17. *(Source: BP Statistical Review of World Energy 2018, June 2018.)*

single that in each of the years within the period considered the production of natural gas increased. The major increase was registered between 2016 and 2017 (3,7%). It is expected that the production of natural gas at world level will continue to grow during the coming decades, as a result of the reduction in the use of crude oil and coal for electricity generation and heating in many countries. Measures adopted by several countries to reduce the negative impact on the environment and population as a result of the use of these two energy sources for electricity generation are responsible for this reduction.

On the other hand, the consumption of natural gas has been increasing during the period 2010—17 as a result of the substitution of crude oil and coal for natural gas for electricity generation and heating in many countries all over the world. The evolution in the consumption of natural gas at world level during the period mentioned above is shown in Fig. 1.8.

Based on the information included in Fig. 1.8, the following can be stated: the consumption of natural gas at world level increased by 15,6% during the period 2010—17, rising from 112,16 billion cubic feet in 2010 to 129,62 billion cubic feet in 2017. It is important to single out that in each of the years of the period considered, the consumption of natural gas at world level increased. It is likely that this trend will continue without change during the coming years, as a result of the measures adopted by several countries all over the world to increase the use of natural gas for electricity generation and heating, and to reduce the use of oil and coal fo this specific purpose.

The world trade in natural gas, both by pipeline and in the form of LNG, is also expected to increase during the coming years, at least until 2040. In the specific case of world LNG trade, it is projected to grow more than double during the period 2012—40, rising from 12 trillion cubic feet in 2012 to 29 trillion cubic feet in 2040; that is to say, the expected growth of 242%.

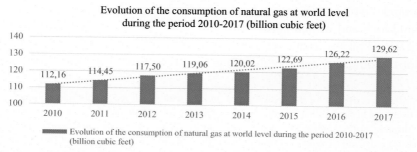

Evolution of the consumption of natural gas at world level during the period 2010-2017 (billion cubic feet)

Evolution of the consumption of natural gas at world level during the period 2010-2017 (billion cubic feet)

Figure 1.8 Evolution in the consumption of natural gas at world level during the period 2010—17. *(Source: BP Statistical Review of World Energy 2018, June 2018.)*

Most of the projected increase in liquefaction capacity is probable to occur in Australia and North America, where many new liquefaction projects are included in the energy sector development plans or are currently under construction. Many of these projects are expected to come online in the next decade. Despite the projection of a steady growth in LNG trade, natural gas flow through oil pipelines is expected to continue to account for most of the world natural gas trade (International Energy Outlook 2016 with Projection to 2040, 2016), for which the construction of new long-distance pipelines and expansion of existing infrastructure until 2040 will be needed. The most significant volumes of natural gas pipeline trade at world level are currently between Canada and the US, and within Europe.

The World Coal Sector

Coal is the third of the three fossil fuels used for electricity generation and heating, and, at the same time, it is the most contaminated of them. Coal has met nearly half of the world's energy demand growth over the past decade. Almost all of this growth was attributable to a rising coal demand from China. It is important to highlight that two-thirds of the world coal consumption is used for electricity generation and heating; the remainder is mainly used for steel manufacturing (Canada's Energy Future 2016. Energy Supply and Demand Projections to 2040, 2016).

According to the International Energy Outlook 2016 with Projection to 2040 report, coal is expected to be the world's slowest-growing energy source during the period 2012—40, rising by an average 0,6% per year, from 153 quadrillions Btu in 2012 to 180 quadrillion Btu in 2040. This foreseeable small increase is considerably lower than the 2,2% per year growth average registered over the past 30 years, and the second-largest energy source worldwide—behind petroleum and other liquids—until 2030.[11] From 2030 through 2040, coal is expected to be the third-largest fossil fuel, next to both liquid fuels and natural gas.

World proved coal reserves are currently sufficient to meet 153 years of global production, roughly three times the R/P ratio for oil and natural gas.

---

[11] "Coal was the first of the fossil fuels to go into widespread use, displacing low-energy firewood as the leading source of fuel in the US, and triggering the country's industrialization in the second half of the 19th century. Within a few decades, the US went from a net importer of coal (mostly from the UK) to a major exporter of this specific fossil fuel, a development made possible by mining the nation's vast reserves of coal" (Oil 2017, 2017).

Proved coal reserves by region in 1997 (%)

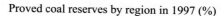

Figure 1.9 Proved coal reserves in 1997. *(Source: BP Statistical Review of World Energy 2018, June 2018.)*

Proved coal reserves by region in 2007 (%)

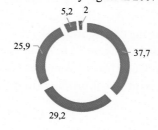

Figure 1.10 Proved coal reserves in 2007. *(Source: BP Statistical Review of World Energy 2018, June 2018.)*

Proved coal reserves by region in 2017 (%)

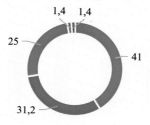

Figure 1.11 Proved coal reserves in 2017. *(Source: BP Statistical Review of World Energy 2018, June 2018.)*

By region, Asia and the Pacific hold the most proved coal reserves (41% of the total), with China accounting for 13,4%. The US remains the largest proved coal reserve holder with 24,2% of the total. North America, including Mexico, has the highest regional R/P ratio (335 years) (BP Statistical Review of World Energy 2018, 2018).

The evolution of proved coal reserves by region in the years 1997, 2007, and 2017 is shown in Figs. 1.9—1.11.

According to Figs. 1.9—1.11, the world proved coal reserves decreased by 6,5% during the period 1997—2017, falling from 1.106.210 million tons in 1997 to 1.035.012 million tons in 2017. According to the BP Statistical Review of World Energy 2018 report, in the specific case of the North America region, proved coal reserves increased by 2,4%, up to 258.709 million tons in 2017 (25% of the total proved coal reserves recorded in that year).

On the other hand, and according to the above-mentioned report, coal production in the North America region decreased by 5,6% during the period 2006—16. During the period 2016—17, coal production increased by 5,9%, as a result of an increase of 6,9% in the US coal production and a decrease of 2% in the Canadian coal production.

The evolution of world coal production during the period 2010—17 is shown in Fig. 1.12.

Considering the data included in Fig. 1.12, the following can be stated: coal production increased by 4,6% during the period 2010—17 rising from 3.601,6 million tons of oil equivalent in 2010 to 3.768,6 million tons of oil equivalent in 2017. However, it is predictable that the production of coal at world level will decrease during the coming years, as

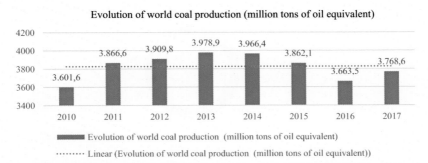

Figure 1.12 Evolution of the world coal production during the period 2010—17. *(Source: BP Statistical Review of World Energy 2018, June 2018.)*

a result of the implementation of the Paris agreement on climatic changes, and the adoption of specific measures to reduce the use of coal for electricity generation and heating in the US, and Canada. The closure of old and inefficient coal-fired power plants used for electricity generation and heating in both countries for environmental reasons is one of the measures that have been adopted by both countries, particularly by Canada.

There is a significant regional variation expected to be registered in the outlook for coal production until 2040. The primary anticipated variations are the following:

- large projected increases for India;
- considerable growths in coal production in Africa and Russia;
- a growth that slows down to decrease gradually after 2025 in China; and
- a few changes in the OECD Europe countries and the US.

However, with the implementation of the Clean Power Plan (CPP), the production of coal in the US is expected to decrease significantly during the period considered (International Energy Outlook 2016 with Projection to 2040, 2016). However, this trend could change if the new energy policy adopted by the current US administration is fully implemented.

Asia and the Pacific are currently the most significant markets for coal trade, consuming, in 2016, a total of 73,8% of the world's coal production. During the period 2018—40, the three largest coal consuming countries is expected to be China, the US, and India, which together is supposed to consume more than 70% of the world's coal production.[12] In 2016, China accounts for almost half of the total coal consumption across the globe (1.887,6 million tons of oil equivalent or 50,6% of the total coal consumption in that year). As a result of a decrease in the growth of China's economy and the adoption of measures to reduce air pollution and the negative impact on the environment and population, the use of coal for electricity generation and heating in China was cut during the last years of the period considered. India is expected to consume 411,9 million tons of oil equivalent or 11% of the world's total, followed by the US with 358,4 million tons of oil equivalent or 9,6% of the world's total.

The evolution of the consumption of coal at world level during the period 2010—17 is shown in Fig. 1.13.

According to Fig. 1.13, the consumption of coal at world level increased by 3,5% during the period considered, rising from 3.605,6 million tons of

[12] In 2016, China, US, and India consumed 71,2% of the world coal consumption.

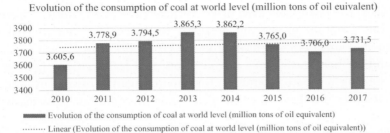

Figure 1.13 Evolution of the consumption of coal at world level during the period 2010—17. (Source: BP Statistical Review of World Energy 2018, June 2018.)

oil equivalent in 2010 to 3.731,5 million tons of oil equivalent in 2017. For a better analysis of the behavior in the consumption of coal at world level, the period considered should be divided into two. During the first period 2010—13, the consumption of coal increased by 7,2% rising from 3.605,6 million tons of oil equivalent in 2010 to 3.865,3 million tons of oil equivalent in 2013. But during the second period 2014—17, the use of coal decreased by 3,4% falling from 3.862,2 million tons of oil equivalent in 2014 to 3.731,5 million tons of oil equivalent in 2017. The peak in the consumption of coal was reached in 2013.

Despite this significant reduction in the consumption of coal for electricity generation and heating registered in the last years, coal still is used to produce around 40% of the world electricity production. However, the negative impact on the environment and the population as a result of the use of coal for electricity generation and heating, and its responsibility with the climatic change that is taking place worldwide is forcing many countries to speed up the transition to the use of cleaner forms of energy for this specific purpose. For this reason, it is likely that the production of coal at world level will continue to decrease during the coming years, mainly due to a significant reduction in the consumption of coal by China, the US, and other countries.

Summing up the following can be stated: worldwide coal consumption is projected to remain roughly the same between 2015 and 2040, with decreasing consumption in China and the US that will offset growth in India. China is expected to continue to be the largest single consumer of coal until 2040, despite a steady decline in the consumption of this type of energy source throughout the period considered. On the other hand, it is also likely that coal consumption in India will continue to grow at an

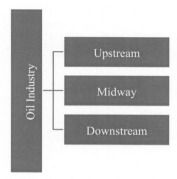

Figure 1.14 The division of the oil industry.

average of 2,6% per year between 2015 and 2040, surpassing the US as the second largest consumer of coal before 2020. In the OECD countries, it is projected that the consumption of coal decreases by an average of 0,6% per year during the period 2015–40, due to the growing competition from natural gas and renewable energies, and a moderate increase in the demand for electricity. Africa, the Middle East, and other non–OECD countries in Asia are predicted to increase the capacity and coal consumption gradually for electricity generation and heating by 2040.

The Crude Oil Industry in the North America Region

The crude oil industry in the North America region is composed of thousands of companies dedicated to the exploration, production, transportation, refining, distribution, and marketing of oil products. This industry is often informally divided into "upstream" (exploration and production), "midway" (transportation and refining) and "downstream" (distribution and commercialization) (see Fig. 1.14). The oil and natural gas have the same upstream sector but different midstream and downstream sectors.

Crude Oil Reserves in the North America Region

Proved crude oil reserves are the estimated quantities that the geological and engineering data indicate that can be a reasonable certainty to be recoverable in the coming years from known reservoirs, under existing technology and current economic and operating conditions. Most of the increases in proved world crude oil reserves since 2000 come from revisions of crude oil reserves already reported in discovered fields, instead of new discoveries. In December 2015, the world's proved crude oil reserves were

Table 1.4 Proved Crude Oil Reserves (millions of barrels).

Proved Crude Oil Reserves (millions of Barrels)		
Country	Rank	Reserves 2017
Venezuela	1	315.878
Saudi Arabia	2	266.455
Canada	3	169.709
Iran	4	158.400
Iraq	5	142.503
Kuwait	6	101.500
UAE	7	97.800
Russia	8	80.000
Libya	9	48.363
United States	10	39.230
World total		**1.726.685**

Note: According to the OPEC database (2017), the US crude oil reserves were reported to be 32.318 million barrels, which is lower than the level of the crude oil reserves reported by the EIA. *Source:* EIA database 2017.

estimated to be 1.656.000 million barrels, which is 2 billion barrels more than the estimate made at the end of 2014 (Xu et al., 2015). In 2017, and according to EIA sources, proved world crude oil reserves were estimated to be 1.726.685 million barrels (see Table 1.4).[13]

In the case of the North America region, its proved crude oil reserves increased, in 2017, by 31,9%. In the same year, the proved crude oil reserves of the US increased by 38,9%, while the proved crude oil reserves in Canada decreased by 7,2%.

About 44% of the world proved crude oil reserves are located in the Middle East. Around 77% are concentrated in eight countries located in different regions including Canada, Iran, Iraq, Kuwait, Russia, Saudi Arabia, United Arab Emirates, and Venezuela. From these countries, only Canada (with oil sands included) and Russia are not members of the OPEC[14] (International Energy Outlook 2016, 2016).

In 2013, the most substantial increase in the world proved crude oil reserves were reported by Venezuela, as it included in its reserves the extra-

---

[13] According to OPEC database 2017, the world proved oil reserves were estimated to be 1.492.164 million barrels.

[14] OPEC: Organization of the Petroleum Exporting Countries. OPEC is a permanent, intergovernmental Organization, created at the Baghdad Conference on September 10–14, 1960, by Iran, Iraq, Kuwait, Saudi Arabia, and Venezuela. OPEC has its headquarters in Vienna, Austria, since September 1, 1965.

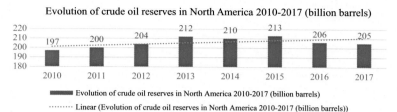

Figure 1.15 Evolution of the proved crude oil reserves in the North America region during the period 2010—17. *(Source: EIA database 2018.)*

heavy crude oil from the Orinoco Belt. As a result, Venezuela's crude oil reserves alone increased by 86 billion barrels. Russia also reported a significant increase in its crude oil reserves by 20 billion barrels in 2013. World crude oil reserves in 2017 are shown in Table 1.4.

On the other hand, it is important to single out that the world's oil shale resources are estimated to contain around 6.050 billion barrels, which makes them 3,5 times the size of the world's conventional crude oil reserves included in Table 1.4. There are over 600 known deposits of oil shale in 33 countries from all continents.

The evolution of the proved crude oil reserves in the North America region during the period 2010—17 is shown in Fig. 1.15.

According to Fig. 1.15, the North America region proved crude oil reserves increased by 4% during the period 2010—17, rising from 197 billion barrels in 2010 to 205 billion barrels in 2017. However, during the period 2010—13, the proved crude oil reserves increased each year for a total of 7,6% but decreased by 3,7% during the period 2013—17 except for 2015. In the period 2015—17, proved crude oil reserves in the North America region declined in each of the years considered. Based on the available information from different sources, it is expected that the proved crude oil reserves in the region will continue to decline at least during the coming years if no new crude oil deposits are found in the area.

Crude Oil Production in the North America Region

The development of crude oil production in the North America region slows down as a result of the drop in crude oil prices registered in the last years, but still, it remains the primary driver of the growth of world crude oil production.

The evolution of the production and consumption of US liquid fuel and their projections during the period 1970—2040 can be found in Fig. 1.16.

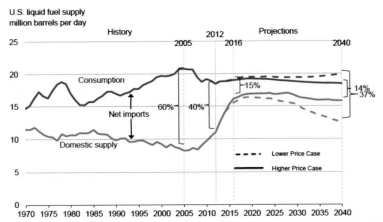

Figure 1.16 Evolution of the production and consumption of US liquid fuel and their projections during the period 1970–2040. *(Source: EIA Preliminary Analysis (Sieminsky, 2015).)*

Fig. 1.16 shows the dependency of the US on crude oil price. Based on the data included in the above chart, the following can be stated: if the crude oil price is low in 2040, then the US net imports in that year could reach 37% of the total crude oil needs of the country, while if the crude oil price is high, then this percentage could fall to 14%; this means a reduction of 23%.

On the other hand, global consumption of liquid fuels is expected to increase from 95 million barrels per day in 2015 to 113 million barrels per day in 2040; this means a growth of 18,9% during the period considered. The non-OECD countries are responsible for most of the rise in oil consumption during the period 2015–40 (39%), while in the OECD countries the use is expected to decrease slightly (3%). The OPEC countries are expected to maintain or even increase their combined market share of crude oil and lease condensates production up to 2040 (International Energy Outlook 2016 with Projection to 2040, 2016).

According to the BP Energy Outlook Focus on North America (2016) report, the liquids supply in the region is expected to expand to 10 million barrels per day by 2035, the highest level ever achieved by this region. This projected growth will occur as a result of the increase in production of about 4 million barrels per day of tight oil in the US, the output of 4 million barrels per day of LNG, and the extraction of two million barrels per day of Canadian oil sands. By 2035, it is likely that the North America region will supply 26% of the liquids worldwide, which will represent an increase of 4% concerning the level currently reported. In the case of the OPEC, its liquid fuels production is probably to stand between 39% and 43% of the total world production throughout the period considered.

The changing pattern of global crude oil supply and demand provokes that regional oil imbalances also change and become more concentrated. In particular, the increase in tight oil production, together with the decrease in demand for crude oil, further reduces North America's dependence on crude oil imports[15] (BP Energy Outlook 2016, 2016). The need for crude oil in the US is expected to decrease by one million barrels per day until 2035, and the US tight oil production is also likely to continue to grow, albeit at a gradually moderate pace. If the increase in Canadian crude oil production is considered, then this will allow the North American region to become a net exporter of crude oil by 2021 at the latest.

According to the World Energy Outlook 2014 report, world production of tight oil is expected to increase during the coming years. The most significant new supplies of tight oil are projected to come from the US, Russia, Canada, and Argentina, which are also beginning to produce tight oil. It is anticipated that these countries will increase their tight oil production during the coming years. However, by 2040, the total production of tight oil in the US is expected to be lower than the tight oil production foreseeable for this country by 2020. But it is supposed to remain higher than the tight oil production anticipated for other countries as can be seen in Fig. 1.17.

It is important to highlight that, between 2008 and 2015, a growth of almost 7 million barrels per day of liquid fuel supply was registered in the North American market. This increase has been only partially offsite at world level by interruptions in the supply of this type of fuel by other oil-producing regions, especially North Africa and the Middle East. Over the past two years, unplanned disruption in crude oil production averaged 3,2 million barrels per day, according to EIA estimates, and amounted to 3,4 million barrels per day in November 2015. Three OPEC countries, Libya, Iraq, and Iran, and two non–OPEC countries, South Sudan and Syria, are responsible for a considerable part of the unplanned interruptions. Global production of liquid fuels in 2014 exceeded consumption and one

---

[15] According to the Medium-Term Oil Market (2015) report, "the North American oil production surge led by unconventional oils—US light tight oil and Canadian oil sands—had produced a global supply shock that would reshape the way oil is transported, stored, refined, and marketed".

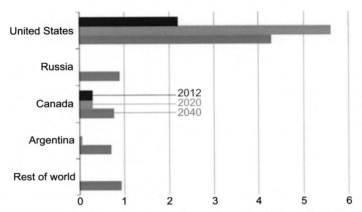

Figure 1.17 World tight oil production. *(Source: International Energy Outlook 2016 with Projection to 2040.)*

year later reached 95 million barrels per day. Excess production was stored increasing OECD inventories up to 2,7 billion barrels in November 2015.

Several factors could provide certain incentives for a significant increase in world liquid production. Among them, the following four elements are of specific relevancy:

- competition among OPEC member countries for an increasing share of the world liquid production market;
- profit requirements of liquid fuels–exporting countries;
- decreasing service costs; and
- the discovery of new and more advanced technology that lowers the production cost and increases recovery rates for tight oil development (International Energy Outlook 2016 with Projection to 2040, 2016).

The evolution of North America crude oil production per day during the period 2011−17 is shown in Fig. 1.18.

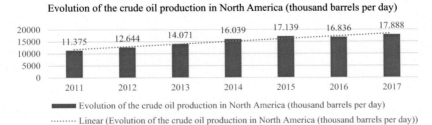

Figure 1.18 Evolution of the production of crude oil per day in North America during the period 2011−17. *(Source: BP Statistical Review of World Energy 2018, June 2018.)*

According to Fig. 1.18, the production of crude oil in the North America region during the whole period increased by 57,3%, rising from 11.375 thousand barrels per day in 2011 to 17.888 thousand barrels per day in 2017. The decrease in crude oil production registered between 2015 and 2016, the first one within the period 2011–17, was a result of the reduction in the US and Canada crude oil production of 5,9% and 6%, respectively. The peak in crude oil production within the North America region during the period considered was reached in 2017. It is expected that the crude oil production in the North America region will increase, at least during the coming years, but at a moderate pace.[16]

According to the Global Trends in Oil and Gas Markets to 2025 report, during the next decade, the North America region will remain the leader regarding growth in production of liquid hydrocarbons. It is also predicted that by 2025 the aggregate volume of liquid hydrocarbon and biofuel production in the US and Canada will reach 19 million barrels per day, thus significantly reducing the region's dependency on oil imports.

Crude Oil Consumption in the North America Region

The consumption of liquid fuels in the North America region is expected to be 22 million barrels per day by 2035, the lowest consumption since 1996. It is important to note that the global demand for liquid fuels has been a critical factor in the world crude oil low prices registered in recent years. In the non-OECD countries, the strong growth in the demand for liquid fuels in the early and middle of the 2000s has been moderating significantly, as a result of the slowdown in the economic growth in key economies such as China, India, and Brazil. The consumption of liquid fuels among the OECD countries, which reached 50 million barrels per day in 2005, has tended to decline since then, reflecting both the growing energy efficiency in the transport sector and the decrease in demand associated with the slow economic growth of key countries.

However, even with those trends tending to reduce demand growth of liquid fuels, it is expected that the global consumption of liquid fuels will increase, as an annual average of 1,1 million barrels per day, according to the International Energy Outlook 2016 with Projection to 2040 report. Based on the information included in that report, the increase in the

---

[16] The greatest increase in crude oil production will come from the deep–water shelf, tight oil reservoirs in the US, and heavy crude oil from Canada and Venezuela.

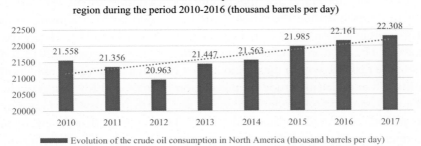

Evolution of the crude oil consumption in the North America region during the period 2010-2016 (thousand barrels per day)

■ Evolution of the crude oil consumption in North America (thousand barrels per day)

········ Linear (Evolution of the crude oil consumption in North America (thousand barrels per day))

Figure 1.19 Evolution of the crude oil consumption in the North America region during the period 2010–17. *(Source: OPEC database 2016 and BP Statistical Review of World Energy 2018, June 2018.)*

consumption of crude oil and other liquids during the whole period 2012—40 is expected to be 34,4%.

By 2025, it is expected that the level of liquids consumption will reach 105 million barrels per day. The most significant demand (90% or more) is likely to come from the transportation sector of developing countries, due to considerable growth in the number of cars as well as for an increase of sea, air, and railway transportation (Global Trends in Oil and Gas Markets to 2025, 2013).

The evolution in the crude oil consumption in the North America region during the period 2010—17 is shown in Fig. 1.19.

For the best analysis of the information included in Fig. 1.19, the period considered can be divided into two: the first one from 2010 to 2012, and the second from 2013 to 2017. During the first period, the consumption of crude oil at a regional level decreased by 2,8%, falling from 21.558 thousand barrels per day in 2010 to 20.963 thousand barrels per day in 2012; in the second period, the regional crude oil consumption increased by 8,7%, rising from 21.447 thousand barrels per day in 2013 to 22.308 thousand barrels per day in 2017. If the whole period is considered, then the local crude oil consumption increased by 3,5%. It is expected that this last trend will continue without change at least during the coming years.

Crude Oil Trade in the North America Region

According to the BP Statistical Review of World Energy 2017 report, North America imported, in 2016, a total of 8.463 thousand barrels per day of crude oil, which represented 20% of the world total crude oil imports per day in that year. The US imported 93% of the total crude oil imported by the region, and Canada the remaining 7%. In 2017, North America imported a total of 8.510 thousand barrels per day of crude oil, which represented 20% of the world total crude oil imports per day in that year. The US imported 92% of the total crude oil imported by the region (1% less than in 2016), and Canada the remaining 8% (1% more than in 2016) (BP Statistical Review of World Energy 2018, 2018).

On the other hand, the North America region exported, in 2016, a total of 3.782 thousand barrels per day of crude oil, which represented 8,9% of the world total crude oil exports per day in that year. Canada exported 87% of the total crude oil shipped by the region. In the case of oil products, the North America region imported, in 2016, a total of 2.838 thousand barrels per day and exported 4.847 thousand barrels per day (BP Statistical Review of World Energy 2017, 2017). In 2017, the North America region exported a total of 4.400 thousand barrels per day of crude oil, which represented 10% of the world total crude oil exports per day in that year. Canada exported 79% of the total crude oil shipped by the region. In the case of oil products, the North America region imported, in 2017, a total of 2.816 thousand barrels per day and exported 4.943 thousand barrels per day (BP Statistical Review of World Energy 2018, 2018).

The Natural Gas Industry in the North America Region

The natural gas industry is a critical segment of the North American economy. It provides one of the cleanest fossil fuels available to all sections of the regional economy, including the generation of electricity and heating. For this reason, since 2015, natural gas has become the primary source of electricity generation and heating in the US. In the case of Canada, natural gas already satisfies 30% of its energy needs, and it is expected that this percentage will increase during the coming years.

It is important to highlight that, in the North America region, the natural gas industry includes exploration, production, processing, transport, storage, and commercialization not only of natural gas but also of its derivatives and by-products. The research and production from natural gas and crude oil form, in many cases, a single industry and, for this reason, many wells produce both oil and natural gas.

Driven by its profitability and the abundant availability of natural gas in the North America region, this type of energy source is ready to undergo a significant transformation, with the aim of becoming the primary fossil fuels used for electricity generation and heating within the region in the next decade. It is important to stress that the long-term demand for natural gas at world level is expected to increase significantly during the coming years, due to the transition to a future economy with the lowest possible CO_2 and other greenhouse gases' emissions, and remarkable growth in the demand for natural gas in the Asia and the Pacific region.

With the increasing use of the hydraulic fracturing (or fracking) technology, shale gas production is now booming in the North America region. Along with sophisticated horizontal drilling equipment that can drill and extract gas from oil shale formations, the use of this new technology not only allows growth in energy supplies but an increase of the North America natural gas reserves. "Despite a domestic supply glut, the fundamentals of natural gas continue to be favorable in the long run, considering the secular shift to the cleaner-burning fuel for power generation globally, and in the Asia and the Pacific region in particular" (Choudhury, 2017).

The EIA forecasts world demand for natural gas to grow during the period 2015—40 from 340 billion cubic feet per day in 2015 to 485 billion cubic feet per day by 2040; this means an increase of 42,6% for the whole period. Countries in Asia and the Middle East—led by China's transition away from coal—will account for most of this increase. On the other hand, the largest producer of natural gas in the world, the US, is expected to take all necessary measures to meet this growing demand, allowing US natural gas companies to sell their production of natural gas at lower prices, compared to the prices that other producing countries could offer. More than 50% of the future growth of the US volume of natural gas will be used to satisfy exports needs but in the form of LNG. According to the International Energy Agency (IEA), it is expected that "the US will vie with Australia and Qatar as the top LNG exporter by 2022" (Choudhury, 2017).

It is well-known that the natural gas market of the North America region is highly integrated. Canada exports its surplus of natural gas to the US, and imports smaller quantities from that country. The increased production of shale gas in the US is displacing Canadian gas imports. Canadian gas exports to the US are now 25% lower than 10 years ago. It is likely that the level of Canada's gas exports to the US will continue to decline in the coming years.

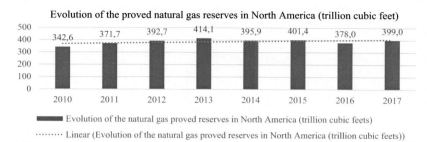

Figure 1.20 Evolution of the proved natural gas reserves in the North America region during the period 2010–17. *(Source: OPEC database 2016 and EIA database 2018.)*

Natural Gas Reserves in the North America Region

According to EIA sources, the world's proved natural gas reserves have grown by about 40% over the past 20 years, reaching a total of 6.950 trillion cubic feet in 2016.[17] In the North America region, the evolution of the proved natural gas reserves during the period 2010–17 is shown in Fig. 1.20.

Based on the data included in Fig. 1.20, the following can be stated: natural gas proved reserves in the North America region increased by 16,7% during the period 2010–17, rising from 342,6 trillion cubic feet in 2010 to 399 trillion cubic feet in 2017. The peak in the proved natural gas reserves was reached, within the period considered, in 2013. It is important to highlight that natural gas proved reserves in the North America region increased each year within the period 2010–13. After that period, natural gas proved reserves began to fall from 414,1 trillion cubic feet in 2013 to 399 trillion cubic feet in 2017; this means a reduction of 3,6%. However, during the last four years, the natural gas proved reserves registered a small (0,8%) but a steady increase in two of the four years period. It is projected that this small percentage increase will continue during the coming years.

According to Figs. 1.21–1.23, the share of North America proved natural gas reserves in the world proved natural gas reserves decreased by 0,7%, during the period considered, falling from 6,3% in 1997 to 5,6% in 2017.

The annual growth rate in the world proved natural gas reserves during the period 1980–95 was significantly higher than the growth registered in more recent years. Since 1995, the annual growth rate in the level of these

---

[17] According to BP sources, at the end of 2018 the world proved natural gas reserves are reported to be 6.831,7 trillion cubic feet. The North America region has 5,5% of this world natural gas reserves.

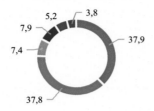

Figure 1.21 Distribution of proved natural gas reserves in 1997. *(Source: BP Statistical Review of World Energy 2018, June 2018.)*

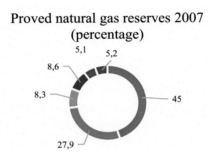

Figure 1.22 Distribution of proved natural gas reserves in 2007. *(Source: BP Statistical Review of World Energy 2018, June 2018.)*

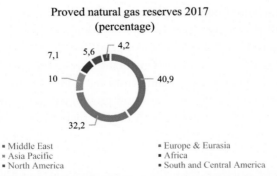

Figure 1.23 Distribution of proved natural gas reserves in 2017. *(Source: BP Statistical Review of World Energy 2018, June 2018.)*

reserves has slowed down at a constant rate of approximately 1,6% per year. In the last 20 years (1996–2016), estimates of proved natural gas reserves increased by 838 trillion cubic feet, or an average of 1,3% per year, compared with 1.179 trillion cubic feet, or 2,2% per year average, during the previous period. Estimated proven natural gas reserves in non–OECD countries increased by 723 trillion cubic feet, or an average of 1,2% per year in the last 10 years, compared to 2,4% per year during the period 1996–2006. Estimates of the world's proved natural gas reserves increased by 31 trillion cubic feet from 2015 to 2016, with more than half of the increase (17 trillion cubic feet) registered by the OECD countries. From 2015 to 2016, proved natural gas reserves in the OECD Americas increased by 29 trillion cubic feet. The most significant change in the proved natural gas reserve estimates was for the US, where proven natural gas reserves increased by 30 trillion cubic feet (9%), rising from 338 trillion cubic feet in 2015 to 369 trillion cubic feet in 2016 (see Table 1.5). The second most

Table 1.5 World Proved Natural Gas Reserves in a Selected Group of Countries as for January 2016.

Country	Proved Natural Reserves (Trillion Cubic Feet)	Percentage of World Total (%)
Russia	1.688	24,3
Iran	1.201	17,3
Qatar	866	12,5
United States	369	5,3
Saudi Arabia	300	4,3
Turkmenistan	265	3,8
United Arab Emirates	215	3,1
Venezuela	198	2,9
Nigeria	180	2,6
China	175	2,5
Algeria	159	2,3
Iraq	112	1,6
Indonesia	102	1,5
Mozambique	100	1,4
Kazakhstan	85	1,2
Egypt	77	1,1
Canada	70	1,0
Norway	68	1,0
Uzbekistan	65	0,9
Kuwait	63	0,9
Rest of the world	591	8,5

Source: International Energy Outlook 2016 with Projection to 2040, 2016. US Energy Information Administration, Office of Energy Analysis, US Department of Energy, IEO 2016.

significant change in proved natural gas reserves during the period 2015—16 was reported by China, where proved natural gas reserves increased by 11 trillion cubic feet (7%), rising from 164 trillion cubic feet in 2015 to 175 trillion cubic feet in 2016 (International Energy Outlook 2016 with Projection to 2040, 2016).

Current estimates of proved natural gas reserves around the world provide a solid basis for supporting market growth through 2040 and beyond (see Table 1.5).

According to Table 1.5, the US is ranked fourth, with proved natural gas reserves at 369 trillion cubic feet (5,3% of the world total); Canada ranked 17th with 70 trillion cubic feet (1% of the world total).

As in the case of other fossil fuel reserves, natural gas reserves are unevenly distributed throughout the world; almost three-quarters of the world's proved natural gas reserves are found in two regions: the Middle East and Eurasia. Despite the high rates of an increase in natural gas consumption, particularly in the last decade, most of the relationships between regional proved natural gas reserves and the level of their production have remained high. Throughout the world, the relationship between proved natural gas reserves and their level of production is estimated at 54 years. By region, these estimates are, according to the International Energy Outlook 2016 with Projection to 2040 report, the following:

- Central and South America: 44 years;
- Russia: 56 years;
- Africa: 70 years;
- Middle East: more than 100 years; and
- US: 13 years.

It is important to note that the US proved natural gas reserves are large enough to allow the continued expansion of both the proved natural gas reserves and their level of production, in most of the economic, technical, and environmental scenarios that can be considered.

It would not be unusual for the reserves-production ratio in a given country to remain constant or grow over time, despite the growth of production. In the last 100 years, there has been a low relation between natural gas reserves and their level of production.

Natural Gas Production in the North America Region

The North America region is expected to increase natural gas production during the period 2012—40 from 30,1 trillion cubic feet in 2012 to 43,9

Table 1.6 Natural Gas Production in the North America Region in 2012 and 2040 (trillion cubic feet).

Country	Natural Gas Production 2012	Natural Gas Production 2040	Total Increase 2012—40
United States	24	35,3	11,3
Canada	6,1	8,6	2,5
Total	30,1	43,9	13,8

Source: International Energy Outlook 2016 with Projection to 2040, 2016. US Energy Information Administration, Office of Energy Analysis, US Department of Energy, IEO 2016.

trillion cubic feet in 2040; this means an increase of 13,8 trillion cubic feet or 45,8% during the period considered (see Table 1.6). In the case of the US, the natural gas production is expected to increase 11,3 trillion cubic feet during the period 2012—40, while in the case of Canada it is expected to grow 2,5 trillion cubic feet. It is important to stress that, in the US, it is likely that the growth in gas production will come mainly from shale gas resources.[18] According to projections included in the International Energy Outlook 2016 report, China, the US, and Russia will account for almost 44% of the overall increase in world natural gas production between 2012 and 2040.

The evolution of natural gas production in the North America region during the period 2012—17 is shown in Fig. 1.24.

According to Fig. 1.24, the evolution of natural gas production in the North America region during the period 2012—17 increased by 13,9%, rising from 28.230,54 billion cubic feet in 2012 to 32.164,60 billion cubic feet in 2017. It is expected that the natural gas production in the North America region will continue to increase during the coming years, particularly in the production of unconventional gas.

On the other hand, the production of tight gas, shale gas, and coalbed methane is expected to increase by 135% during the period 2012—40, rising from 20 trillion cubic feet in 2012 to 47 trillion cubic feet in 2040. "However, numerous uncertainties could affect future production of those resources. There is still considerable variation among estimates of

[18] In the US, natural gas production increased by 1.687,34 billion cubic feet in 2014 and by 1.168,43 billion cubic feet in 2015. However, in 2016, natural gas production fell in 610,69 billion cubic feet. "This is the first annual decrease since the beginning of the shale gas revolution" (Natural Gas Information: Overview, 2017). According to EIA database 2018, the US production of dry natural gas in 2017 reached 26.854.288 million cubic feet, and 28.814.028 million cubic feet in marketed production.

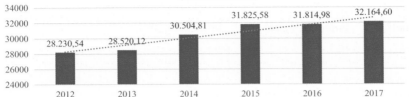

Evolution of the natural gas production in North America
(billion cubic feet)

Figure 1.24 Evolution of natural gas production in North America. *(Source: BP Statistical Review of World Energy 2018, June 2018.)*

recoverable shale gas resources in the US and Canada, and estimates of recoverable tight gas, shale gas, and coalbed methane for the rest of the world are more uncertain, given the sparse data currently available" (International Energy Outlook 2016 with Projection to 2040, 2016).

The US production of shale gas is expected to grow by 92,3% during the period 2012−40, rising from 13 trillion cubic feet in 2012 to 25 trillion cubic feet in 2040, an increase of 12 trillion cubic feet within the period considered, counteracting the decrease in the production of shale gas from other sources. In 2040, shale gas is expected to represent almost 70% of the total US gas production, according to the International Energy Outlook 2017 report.

In Canada, future gas production is expected to come mainly as tight and shale gas from two regions. These regions are British Columbia and Alberta. According to different experts' opinions, it is estimated that tight gas production will represent 15,8% of the total gas production in Canada. The offshore production will represent 8% of that total. An estimate of 6,2% of this production is projected to come from coalbed methane obtained by the extraction of coal, as well as from other onshore resources.

Canada ranked fifth in the production of dry natural gas, despite having very limited proved natural gas reserves, and is the fourth largest net exporter of natural gas next to Russia, Qatar, and Norway. Although Canada has plans to export LNG to other countries, all of Canada's current exports of natural gas are shipped to the US markets through gas pipelines.[19] In Canada, it is expected that much of the growth in natural gas production by 2040 will come from increased production of tight gas and some shale gas production from British Columbia and Alberta. In the case of the US, the

---

[19] In 2016, pipeline gas imports to the US from Canada increased 10,9% (Natural Gas Information: Overview, 2017).

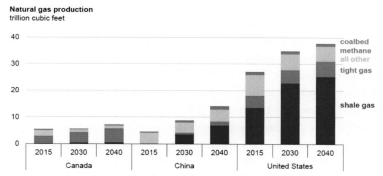

Figure 1.25 Natural gas production in the US and Canada during the period 2015—40. *(Source: International Energy Outlook 2017, September 2017. Energy Information Administration (EIA).)*

significant increase in the production of natural gas will come, in 2040, from shale gas (see Fig. 1.25). Shale gas in the US and tight gas in Canada are now and are expected to be the main components in the production of natural gas during the whole period considered.

Natural Gas Consumption in the North America Region

The demand for natural gas in the North America region is estimated to grow by 1,4% annually, which is 0,4% lower than the estimated annual world increase, making it the only fossil fuel to rise its demand within the region during the period 2014—35.[20] This growth is helped by a greater supply of natural gas, particularly shale and tight gas, and the adoption of environmental policies to support the use of this type of energy sources, especially for electricity generation and heating. The purpose of these measures is the reduction of the current negative impact that the use of certain fossil fuels have on the environment and the population, by replacing the use of coal and oil for electricity generation and heating. The consumption of natural gas and renewable energy sources in the North America region is expected to displace coal in electricity generation and heating by 2023. The share of natural gas in electricity generation and heating is projected to reach 36% by 2035, compared to the current 22%; that represents an increase of 14% for the whole period considered.

---

[20] In 2016, an increase in 1,1% in the demand of natural gas was registered in OECD Americas, mainly driven by the growth in the US (409,48 billion cubic feet), where power generated from natural gas also increased (3,4% electricity generated from natural gas, based on annual electricity data) (Natural Gas Information: Overview, 2017).

An increase is also expected in the use of natural gas in the industry with the aim of reaching a market share of 46% by 2035.

On the other hand, the demand for natural gas in the US is expected to grow almost 21 trillion cubic feet per day by 2035, with power generation by 14 trillion cubic feet per day; in the industry by 5 trillion cubic feet per day; and in transportation by 3 trillion cubic feet per day. In other sectors, a decrease is expected in the use of this type of energy source of at least 1 trillion cubic feet per day. Finally, the demand for natural gas in Canada is expected to increase by 8,6 trillion cubic feet, driven mainly by power generation and industry (BP Energy Outlook Focus in North America, 2016; International Energy Outlook 2016 with Projection to 2040, 2016).

It is important to highlight that the consumption of natural gas from the US would be slightly lower if the Trump administration implements the CPP approved by the Obama administration. But even in this case, it will continue to be an attractive fossil fuel for electric and industrial power generation, not only in the North American region but other regions as well. In the specific case of electricity generation, the use of natural gas as a fuel is an attractive option for new natural gas-fired power plants to be built in the future within the region, due to low capital costs, favorable heat rates, relatively low fuel cost, and more flexibility in its operation. It is expected that the use of natural gas for electricity and industrial power generation will represent almost 75% of the projected increase in total natural gas consumption between 2015 and 2040 (see Fig. 1.26). During the period considered, the use of natural gas in the transportation sector will increase slightly.

Finally, the evolution of natural gas consumption in all sectors in the North America region during the period 2012—17 is shown in Fig. 1.27.

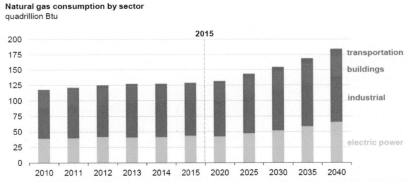

Figure 1.26 Natural gas consumption by sectors and by year during the period 2010—40. *(Source: International Energy Outlook 2017, September 2017. Energy Information Administration (EIA).)*

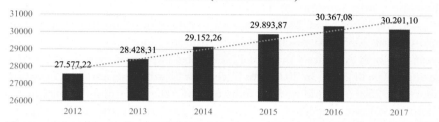

Figure 1.27 Evolution of natural gas consumption in the North America region during the period 2012–17. *(Source: BP Statistical Review of World Energy 2018, June 2018.)*

By the data included in Fig. 1.27, the following can be stated: the consumption of natural gas in the North America region increased by 9,5% during the period 2012–17, rising from 27.577,22 billion cubic feet in 2012 to 30.201,10 billion cubic feet in 2017. For the first time during the period under consideration, the consumption of natural gas in the region registered a decrease of 0,6%. However, it is expected that the consumption of natural gas in the North America region will continue to increase during the coming years.[21]

Exports of Natural Gas and Liquid Natural Gas in the North America Region

The North America region is projected to become a significant exporter of natural gas by 2020, although natural gas flows from Russia to Europe and Asia are expected to show the highest growth. In that year, however, the most substantial volumes of natural gas marketed internationally through pipelines are supposed to be reported between Canada and the US.[22]

According to EIA sources, the natural gas trade between Canada and the US is very high: Canada is not only the most significant partner for energy trade with the US but is also the largest source of US natural gas

---

[21] In 2015, the US was the top country by natural gas consumption in the world, followed closely by Russia. In 2017, the total annual consumption of natural gas by the US reached 27.090.166 million cubic feet.

[22] According to Today in Energy 2017 report entitled "Canada is the United States' Largest Partner for Energy Trade", "Canada is the largest energy trading partner of the United States. Based on the latest annual data from the US Census Bureau, energy accounted for about 5% of the value of all US exports to Canada and more than 19% of the value of all US imports from Canada in 2016."

imports—mostly through pipelines that cross Idaho and Montana. Mexico is the largest destination of US natural gas exports—primarily through pipeline border crossings in Texas.

On the other hand, North American LNG exports are also expected to grow until 2035 and reach almost 18 trillion cubic feet per day by that year.[23] North America's share of LNG global exports is expected to reach around 30% in 2030. In 2035, it is likely that North American LNG exports represent 25% of world trade, a reduction of 5% concerning the maximum exports volume expected in 2030. The growth in the LNG exports coincides with a significant change in the regional trade pattern. The US is expected to become a net exporter of natural gas before the next decade, with net exports growing to 5,6 trillion cubic feet by 2040, while the dependence of Europe and China on imported gas is expected to continue to rise. The US, backed by a 135% growth in shale gas production, will become a net exporter of shale gas in 2019 and it is expected to export more than 20 trillion cubic feet per day by 2035 (BP Energy Outlook Focus in North America 2016, 2016).

Finally, the evolution of the exports of dry natural gas in the North America region during the period 2010—17 is shown in Fig. 1.28.

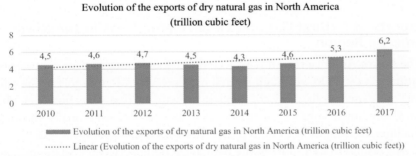

Figure 1.28 Evolution of the exports of dry natural gas in the North America region during the period 2010—17. *(Source: EIA and Statista databases.)*

[23] According to the Canada's Energy Future 2013 report, it is expected that during the period 2019—23, an increase in LNG exports will be registered (from 1 billion cubic feet per day in 2019, to 2 billion cubic feet per day in 2021, and to 3 billion cubic feet per day in 2023). In other words, during the period 2019—23, the Canadian export of LNG is expected to increase three times. Following this trend, in 2035, it is expected that the LNG production in Canada will reach 4,5 billion cubic feet per day, an increase of 50% with respect to the level expected in 2023.

Based on the data included in Fig. 1.28, the following can be stated: the exports of dry natural gas in the North America region increased by 37,8% during the whole period considered, rising from 4,5 trillion cubic feet in 2010 to 6,2 trillion cubic feet in 2017. It is projected that the export of dry natural gas within and by the North America region will continue to increase during the coming years, mainly to countries outside the area.

Imports of Natural Gas and Liquid Natural Gas in the North America Region

The US Natural Gas Imports and Exports 2016 (2017) report single out that, in 2016, natural gas imports increased in the US by 10% with respect to the level registered in 2015, reaching the amount of 3.000,7 billion cubic feet. A total of 1,8 billion cubic feet of natural gas imports come by pipeline from Canada.[24] In turn, in 2016, Canadian imports of natural gas increased by 11% reaching the amount of 2.912 billion cubic feet. "Natural gas imports into Canada from the US had been increasing as shale gas produced in the Northeast US entered the Ontario market. Imports of natural gas have declined since peaking in 2011" (Canada's Energy Future 2016, 2016).

On the other hand, imports of LNG by the US decreased by 3%, reaching the amount of 88,4 billion cubic feet in 2016, the second-lowest level since 1998. The US significant LNG imports come from Trinidad and Tobago.[25] It is important to highlight that the US "also imported a small amount of compressed natural gas from Canada in 2016, almost the same level reported in 2014 and 2015" (US Natural Gas Import and Exports 2016, EIA 2016). A total of 0,06 billion cubic feet or 3% of total natural gas imports by Canada was imported as LNG from the US.

In summary, the following can be stated: in 2016, the total natural gas imports by the North America region was 5.912,1 billion cubic feet. The LNG imports by the North America region was 0,31 billion cubic feet per day (Energy Fact Book 2016–17, 2016).

In the specific case of the US, the evolution of the US gas imports during the period 2011–16 is shown in Table 1.7.

---

[24] During the period 2011–15, the total US natural gas imports average was 2.980,7 billion cubic feet, 0,7% lower than the natural gas imports registered in 2016. "Pipeline imports to the United States from Canada increased 10,9%" (Natural Gas Information: Overview 2017, 2017).

[25] In 2015, the structure of the imports by the US of LNG was the following: 85% came from Trinidad and Tobago, 14,8% from Spain, and 0,2% from the US by truck" according to Canadian National Energy Board and other sources.

Table 1.7 Evolution of the US Gas Imports During the Period 2011—16.

	Five Years Average (2011—15)	2015	2016	2016 Versus 2015 (%)
Pipeline	2.825	2.626,3	2.912	11
LNG	154,2	91,5	88,4	−3
CNG	0,3	0,3	0,3	3
Total	2.979,5	2.718,1	3.000,7	10

Source: Natural Gas Imports and Exports, 2016. US Energy Information Administration (EIA).

Finally, regarding the imports of gas by the North America region, the following can be stated: "the North America natural gas market is backed by a mature and well-integrated physical and financial market structure and substantial domestic natural gas production. Imports have historically faced competition from domestic supplies, which are priced in established trading hubs based on the supply and demand of natural gas in the region" (World Energy Resources 2016, used with the permission of the World Energy Council, www.worldenergy.org).

The Unconventional Energy Revolution in the North America Region

The unconventional oil and gas revolution—shale gas and what has become to be known as "tight oil"—is one of the most crucial energy changes reported so far in the 21st century.[26] Because of this revolution, today more emphasis is placed on energy innovation than ever before. Without a doubt, there is excellent potential in the use of shale gas and tight oil in different economic sectors in Canada as well as in the US.

According to Crude Oil Facts (2018), tight oil resources are found primarily in a belt that stretches from central Alberta to southern Texas. The Permian (mainly in West Texas), the Bakken (North Dakota, Montana, Saskatchewan, and Manitoba), and the Eagle Ford (South Texas) tight oil formations are the most significant sources of tight oil production in North America. Prospective resources have also been identified throughout the Rocky Mountain region, the US Gulf Coast, and the northeastern US/Eastern Canada (including Anticosti Island and Western Newfoundland and Labrador).

---

[26] "Shale and other new sources of oil, like oil sands currently being developed in Canada, offer important new North American energy supply options" (Oil 2017, 2017).

Table 1.8 Technically Recoverable Shale Gas Reserves in the North America Region.

Country	Trillion Cubic Feet
United States	621,54
Canada	572,10
North America	1.223,64

Source: World Energy Resources (2016) (used with the permission of the World Energy Council, www. worldenergy.org).

Undoubtedly, the unconventional energy revolution in North America has changed global energy flows. For example, now North America imports less energy from other regions than before and, at the same time, increased the supply of LNG to the Asian markets. At this moment, there is an increase in the export of US coal to Europe and Asia's markets, since natural gas has replaced this specific type of energy source for electricity generation and heating in the US. The North American technically recoverable shale gas reserves are shown in Table 1.8.

The US and Canada are today the only countries in the world that produce gas and oil from shale formations on a commercial scale.[27] In the specific case of Canada, "future natural gas production is expected to come mainly from tight resources, from several regions in British Columbia and Alberta" (International Energy Outlook 2017, 2017). Other countries have opened exploratory test wells, but for noncommercial purposes. China has just begun the commercial production of oil and gas from shale formation located within its territory. Shale gas production is projected "to account for nearly 50% of China's natural gas production by 2040, making the country the world's largest shale gas producer after the US" (International Energy Outlook 2017, 2017). Global shale gas is expected to grow by 5,6% per year during the period 2014—35, much more than the growth of total natural gas production. As a result, the share of shale gas in world natural gas production is expected to increase by 13% (from 11% in 2014 to 24% by 2035).

Global growth in shale gas supply is dominated by the North American production,[28] as it has been over the past decade. Today shale gas supply

---

[27] Shale resource development accounts for 50% of the US natural gas production in 2015. It is expected that, by 2040, nearly 70% of the natural gas production in the US will be shale gas "as the country leverages advances in horizontal drilling and hydraulic fracturing techniques and taps into newly discovered technically recoverable reserves" (International Energy Outlook 2017, 2017).

[28] In the first half of the period 2014—35, almost all of the growth in shale gas output stems from the US. However, by 2035 "China will become the largest contributor to growth in shale gas production" (BP Energy Outlook 2016, 2016).

accounts for about two-thirds of the increase in global supplies of this type of natural gas. It is important to stress that during the period considered shale gas is expected to expand significantly outside of the North American region, especially in China. In other countries in Asia and the Pacific region, the shale gas production is expected to reach 13 billion of cubic feet per day by 2035. For the last 10 years of the period 2014–35, it is projected that about half of the increase in shale gas supply will come from outside the North American region. According to the BP Energy Outlook 2016 report, Asia and the Pacific region are expected to have a shale gas production of more than 10% of global shale gas production by 2035.

It is important to highlight that during the coming years, "US natural gas exports (other than gas transported by pipelines to Canada and Mexico) would be in the form of LNG" (Erbach, 2014). It is indispensable to convert the current LNG import terminals (which had been built with the expectation of increasing gas imports) into LNG export terminals to guarantee the exports of natural gas by the US during the coming years. Without a doubt, the experience of the North America region in the development of unconventional energy sources can serve as an example for the development of this alternative energy resource in other areas around the world, with a significant impact on its use in different economic sectors (Erbach, 2014).

It is also essential to stress that the shale oil revolution in the US has been possible due to geological, geographic, industrial, financial, and regulatory factors that exist explicitly in the North American region. Over the next few years, it will be confirmed if the shale oil revolution can be replicated in other countries and areas around the world.

Undoubtedly, the shale oil revolution has a significant economic impact and is responsible for the fall in natural gas prices and in the reduction of energy imports by the US. Low natural gas prices have benefited households and industry, especially the steel production industry, as well as the production of fertilizers, plastics, and basic petrochemicals. According to EIA sources, the tight oil production is likely to increase from 4,98 million barrels per day in 2015 to 10,36 million barrels per day in 2040; this represents an increase of 208%.

On the other hand, it is important to highlight that the production of tight oil is expensive, and for this reason, a high oil price is required with the aim of ensuring that its extraction and subsequent commercialization can be economically viable. According to different analysts' opinions, they do not expect additional production capacity to lead to lower oil prices, but it may well prevent oil prices from rising further during the next years.

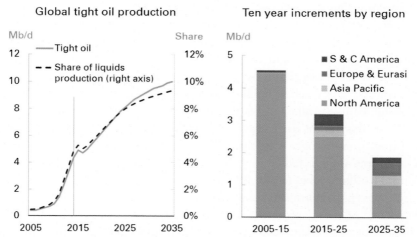

Figure 1.29 Global tight oil production. *(Source: BP Energy Outlook 2016 Edition. Outlook to 2035, 2016.)*

US shale gas, tight oil elsewhere, Brazilian oil deep-water, and Canadian oil sand and biofuels, all together are projected to "grow by 16 million barrels per day, accounting for around half of non-OPEC production in 2035" (BP Energy Outlook, 2016). Without a doubt, the most important unconventional fossil fuels for the US are shale gas and tight oil, produced by horizontal drilling and hydraulic fracturing (fracking) (Erbach, 2014).

According to Fig. 1.29, the production of tight oil in the North America region in comparison to other areas around the world is expected to decline sharply from around 4,5 million barrels per day during the period 2005–15 to approximately 1 million barrels per day during the period 2025–35. Despite this decrease, the North America region is expected to continue to lead the production of tight oil at world level during the period 2025–35. A significant increase in the production of tight oil is likely to occur in Europe and Eurasia, followed by the Asia Pacific region. Analysts estimate that, in 2015 only, US$150 billion were spent in the development of North American tight oil fields. Undoubtedly, the substantial increase in tight oil production is one of the causes behind the reduction in oil price registered at the end of 2014 (Ovale, 2014).

The energy trade between the US and Canada amounted to almost US$67 billion in 2016. However, despite the expected increase of world tight oil production in 5,7 million barrels per day during the period considered, tight oil production is expected to reach less than 10% of all liquid's output in 2035.

Summing up the following can be stated:

- It has been estimated that the global production of shale gas in North America will represent 77% of global shale gas production by 2035 and 84% of global tight oil production.
- Natural gas production is expected to increase until 2035, with tight gas, shale gas, and coalbed methane production from the OECD countries rising from 20 trillion cubic feet in 2012 to 47 trillion of cubic feet in 2040; this means an increase of 135%. During the same period, the production of tight gas, shale gas, and coalbed methane from other countries is expected to grow from almost 2 trillion cubic feet in 2012 to 34 trillion cubic feet in 2040; this means an increase of 17 times.

However, numerous uncertainties could affect the future production of these types of energy resources. There is still considerable variation between estimates of recoverable shale gas resources in the US and Canada, and the forecast of tight gas, shale gas, and coalbed methane for the rest of the world are more uncertain, given the limited data currently available.

The Coal Industry in the North America Region

The coal industry in the North America region has a very uncertain destiny for the next few decades due to environmental considerations and despite efforts made by the current US administration to protect the US coal industry. Thermal coal is facing a long-term slow decline, and the metallurgical coal price is in a nonrecovery status. This situation is not expected to change at least during the coming months. For this reason, only a minimal number of new coal-fired power plants was built in the US after 1990. As a consequence of this situation, the US has today an aging fleet of coal-fired power plants with many small, inefficient generating units lacking modern pollution controls (Tierney, 2016).

On the other hand, US and Canadian coal producers are facing the slowdown in steel production rates, mainly in China, along with the global steel overcapacity that exists in the North America region. Also, the US thermal coal industry is fighting against the low prices of natural gas. Undoubtedly, international and national environmental regulations will put downward pressure on domestic consumption of coal in the foreseeable future, particularly in the North America region and China, and the new investment of energy capacity will be directed toward the increase in the use of natural gas and renewable energies for electricity generation and heating in both cases.

Finally, it is important to highlight that market forces have influenced the US coal industry for decades, and it is expected to continue to do so in the future. In the last quarter of the 20th century, the US economy grew, as well as the standard of living of its population, as the nation became electrified. This economic growth came about thanks to the vast resources of cheap coal available throughout the country and the fact that the coal companies produced increasingly large quantities of this type of energy source for marketing. As a result of the measures adopted, particularly "productivity improvement, coal mining output, and growing share of power production, the US coal industry looked to be reasonably strong as of the year 2000" (Tierney, 2016).

Coal Reserves in the North America Region

The total proved coal reserves in the North America region at the end of 2017 can be found in Table 1.9.

As of January 1, 2016, about 18,3 billion short tons of recoverable coal reserves were located at the US production mines, which is a meager amount in comparison with the level of coal reserves reported by US authorities in that year. The main reason for this vast variance is because the amount of coal reserves in the production mines is a small portion of the total amount of coal that is estimated to exist within the whole US territory. It is evident that the total coal resources existing throughout the country are difficult to predict with high precision because they are buried underground. However, according to estimates given by the EIA, total carbon reserves (including undiscovered coal) in the US could reach 3,9 trillion short tons. By the current EIA data, the US has some 251 billion short tons of recoverable coal reserves (24,2% of world coal reserves) and Canada 6,6

Table 1.9 Total Coal Reserves in the North America Region at the End of 2017 (million short tons).

Country	Anthracite and Bituminous	Subbituminous and Lignite	Total	Share of World Total (%)	R/P Ratio
United States	220.800	30.116	250.916	24,2	357
Canada	4.346	2.236	6.582	0,6	111
North America*	226.306	32.403	258.709	25	355

Note: *indicates that the value includes the coal reserves in Mexico also.
Source: BP Statistical Review of World Energy 2018, June 2018.

Figure 1.30 US coal resources per region. *(Source: EIA database 2017.)*

billion short tons (0,6% of world coal reserves).[29] In 2015, coal production was approximately 0,9 billion short tons. With this level of production, recoverable coal reserves would satisfy current market demand within the North America region for around 287 years.

In 2016, six US states had 77% of the total US proved coal reserves, and another 26 states had the remaining 23%. These six states are Montana, with 25%; Illinois, with 22%; Wyoming, with 12%; and West Virginia, Kentucky, and Pennsylvania with 6% each. The three regions with the largest coal reserves are the following: Western, Interior, and the Appalachian areas (see Fig. 1.30).

On the other hand, Canada has abundant coal reserves throughout its territory, with large amounts of metallurgical and thermal coal that have an apparent commercial character and are currently exploited in British Columbia, Alberta, Saskatchewan, and Nova Scotia. Significant coal formations have also been founded in Yukon, Ontario, Newfoundland and Labrador, the Northwest Territories, and Nunavut. However, the resources located in these regions are not exploited yet.

At the end of 2017, Canada had 6.582 million short tons of recoverable coal reserves or approximately 111 years of exploitation at the current

---

[29] According to the World Energy Resources 2016 report, the proved coal reserves in the US in 2016 were 237.295 million short tons and in Canada were 6.582 million short tons.

production rate (60,4 million short tons).[30] These coal reserves represent only 0,6% of the world proved coal reserves.[31]

According to Table 1.10, the US is the country with the highest coal reserves followed by Russia, China, India, and Australia.

Proved world coal reserves at the end of 2017 by type of coal and by regions are shown in Table 1.11.

According to Table 1.11, the North America region has, in 2017, the third significant proved world coal reserves (22,8%) with a total of 258.709 million short tons. The proved coal reserves in the North America region was 6,1% higher than the level registered in 2015.

Coal Production in the North America Region

After gaining share since 2000, the world growth of coal is projected to slow sharply, 0,5% per year during the coming years, due to the adoption, at international, regional, and national levels, of a group of measures to reduce the negative impact that the use of coal for electricity generation and heating is causing on the environment and the population. It is anticipated that, by 2035, the world share of coal in primary energy is at an all-time low stage, with natural gas replacing it as the world's second largest fuel source, particularly for electricity generation and heating.

The evolution of coal production within the North America region during the period 2010—17 is shown in Fig. 1.31.

According to Fig. 1.31, the production of coal in the North America region during the period 2010—17 decreased by 28,1%, falling from 559,1 million tons of oil equivalent in 2010 to 402,4 million tons of oil equivalent in 2017. It is expected that the production of coal and its use as a fuel for electricity generation and heating in the North America region will continue to decline during the coming years. This trend is due to several measures adopted by the governments of the US and Canada to reduce the negative impact on the environment and population as a result of the use of coal for electricity generation and heating in both countries.

As shown in Table 1.12, coal production in the OECD Americas and the US will be significantly reduced, if the current US administration definitively implements the proposed CPP.

---

[30] According to Canadian sources, out of the 60,4 million short production per year registered, more than 34,5 million tons is metallurgical grade coal, used during the manufacture of steel.

[31] In 2016, this percentage reached 1%.

Table 1.10 World Proved Coal Reserves by Countries and Regions in 2014, 2015, and 2017 (million tons).

Rank	Country/Region	Coal Reserves in 2014	Rank	Country/Region	Coal Reserves in 2015	Rank	Country/Region	Coal Reserves in 2017
1	United States	253.124	1	United States	237.295	1	United States	250.916
2	Russia	160.364	2	Russia	157.010	2	Russia	160.364
3	China	131.614	3	China	114.500	3	Australia	144.818
4	Australia	106.259	4	Australia	76.400	4	China	138.819
5	India	90.276	5	India	60.600	5	India	97.728
14	Canada	6.582	12	Canada	6.582	14	Canada	6.582
	North America	259.706		North America	243.880		North America (including Mexico)	258.709
	World total	984.624		World total	891.531		World total	1.035.012

Source: World Energy Council and BP Statistical Review of World Energy 2018, June 2018.

Table 1.11 Total Proved World Coal Reserves by Type of Coal and by Regions at the End of 2017 (million short tons).

Region	Anthracite and Bituminous	Subbituminous and Lignite	Total	Share of Total	R/P Ratio
Asia Pacific	314.325	109.909	424.234	41%	79
Europe and Eurasia	154.382	169.251	323.633	31,2%	284
North America	226.306	32.403	258.709	25%	335
Middle East and Africa	14.354	66	14.420	1,4%	53
South and Central America	8.943	5.073	14.016	1,4%	141
World	**718.310**	**316.702**	**1.035.012**	**100%**	**134**

Note: Reserves-to-production (R/P) ratio: the production divides the reserves remaining at the end of any year in that year, the result is the length of time that those remaining reserves would last if the production were to continue at the rate.
Source: BP Statistical Review of World Energy 2018, June 2018.

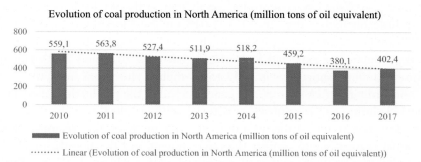

Evolution of coal production in North America (million tons of oil equivalent)

▬▬▬ Evolution of coal production in North America (million tons of oil equivalent)

·········· Linear (Evolution of coal production in North America (million tons of oil equivalent))

Figure 1.31 Evolution of coal production in the North America region during the period 2010—17. *(Source: BP Statistical Review of World Energy 2018, June 2018.)*

Considering the information included in Table 1.12, it is expected that the coal production in the North America region decreases by only 0,8% during the period 2012—40, falling from 1.089 million short tons in 2012 to 1.081 million short tons in 2040 if the US government does not implement the CPP. If the CPP is applied, then the coal production in the North America region is expected to decrease significantly (12,5%), falling from 1.089 million short tons in 2012 to 953 million short tons in 2040. In the

Table 1.12 World Coal Production by the OECD Americas, the US, and Canada During the Period 2012—40 (million short tons).

Region/ Country	2012	2020	2025	2030	2035	2040	Average Annual Percent Change
OECD	2.237	2.341	2.396	2.366	2.356	2.351	0,2
OECD with CPP	2.237	2.150	2.092	2.124	2.154	2.223	0,0
OECD Americas	1.107	1.125	1.143	1.122	1.122	1.096	0,0
OECD Americas with CPP	1.107	934	839	880	920	968	−0,5
United States	1.016	1.044	1.060	1.048	1.045	1.020	0,0
United States with CPP	1.016	853	756	806	843	892	−0,5
Canada	73	67	70	60	63	61	−0,6

Source: EIA database 2016.

specific case of the US, coal production in 2040 without CPP is expected to be a little bit higher than in 2012 (1.016 million short tons in 2012 and 1.020 million short tons in 2040). In the case of Canada, the reduction in coal production is expected to be 16,5% between 2012 and 2040, falling from 73 million short tons in 2012 to 61 million short tons in 2040.

According to EIA sources, the general trend in the production of coal in the US during the coming years is to be significantly lower than in past decades, as a result of the adoption of strict international, regional, and national regulations to protect the environment and the population. EIA estimates that by 2025, coal production in the US could reach 15,4% of the total US energy production, compared to approximately 19,8% achieved in 2015 (slightly less than 18% if the CPP above is not implemented) (Tierney, 2016). The US stood second to China in the production of coal in 2016, with 10% of the world coal production.

In the case of Canada, coal has been used as an energy source since the 18th century. Canada is in the top 10 of the list of countries with the most extensive world coal reserves. Nonetheless, Canada is not a significant world coal producer, and, for this reason, it is ranked 12th according to Natural Resources Canada (see Table 1.13). Canada coal production

Table 1.13 World Coal Production in 2017.

Rank	Country	Percentage (%)
1	China	46,4
2	United States	9,9
3	Australia	7,9
4	India	7,8
5	Indonesia	7,2
10	Canada	0,8

Source: BP Statistical Review of World Energy 2018, June 2018.

was 1% of the total world production in 2016. The country also produces more coal than it consumes and due to the growing global demand for this type of energy source for metallurgical activities, almost half of its production is exported, mainly to Asian countries where the need for metallurgical coal is high.

Canada's coal production peaked in 1997. In this year the country produced approximately 78 million short tons. In the last 10 years, Canada's coal production has remained stable at 73 million short tons of coal per year (Quigley, 2016). This production is almost evenly divided between coking coal used in steel production and thermal coal, which is used to produce electricity. As a result of the energy measures adopted by the Canadian government, it is expected that during the period 2020–40, coal production in Canada will fall from 67 million short tons in 2020 to 61 million short tons in 2040, a decrease of 9% (see Table 1.12). This trend in the reduction in the use of coal for electricity generation and heating is likely to be the same after 2040.

The use of thermal coal for electricity generation and heating foresees a decrease in the demand of this type of fossil fuel until the year 2030, when the government of Canada expects to eliminate the use of coal-fired power plants for this specific purpose. On the other hand, and according to several experts' forecast, it is very likely that the production of coking coal by Canada will remain stable during the coming years, despite the fact that the Canadian coal industry has been targeted by the environmental movement for its negative impact on the environment, on the people's health, and climatic change. Currently, 61,9 TWh or 9,6% of the electricity in Canada is generated by coal-fired power plants. In 2040, it is expected that only 4,1 TWh or 0,6% of the total electricity will be produced in Canada by coal-fired power plants (Canada's Energy Future 2017. Energy Supply and Demand Projections to 2040, 2017).

Coal Consumption in the North America Region

The world currently consumes more than 7,7 billion short tons of coal, which is used by a variety of economic sectors, including power generation, iron, and steel production, cement manufacturing, among others. Despite international efforts to reduce the use of coal for electricity generation and heating, coal currently generates 40% of the world's electricity[32] and is expected to continue to have strategic participation in this crucial sector, at least in several countries, during the next three decades.

It is expected that the North America coal consumption will decrease by 3,3% per year during the period 2014–35, reaching, at the end of this period, its lowest level recorded until today (19% in 2035 from 44% today; this means a reduction of 25%).[33] For this reason, the region is projected to reduce CO_2 emissions by 0,4% per year during the period considered compared to the 0,3% annual growth registered in the last 20 years.

However, the switch to a low-carbon energy sector in the North America region will not happen fast enough to keep global warming below 2°C (3,6 degrees Fahrenheit), the most ambitious goal adopted by the Paris Climate Agreement approved in 2016. An additional investment of US$5,3 billion is required in the energy sector by 2040, in addition to the US$7,8 billion already projected for that year, to achieve the above objective.

On the other hand, the coal demand is projected to decline within the North America region by 51% over the period 2014–35, despite a 7% increase in the overall power demand, reaching the lowest level since 1965. For this reason, the carbon emission is expected to decline 8% during the same period (BP Energy Outlook Focus in North America, 2016).[34] In the specific case of the US, a preliminary analysis of the CPP shows potential reductions of 21% in US coal consumption in 2020 and 24% in 2040, in comparison with the projection included in the International Energy Outlook 2016 with Projection to 2040 report.

---

[32] In the US, in 2000, more than 90% of the country coal supply was used for electricity generation and coal-fired power plants generated 53% of all electricity produced in the country. In 2015, coal-fired power plants generated only 33% of the total electricity produced by the US, a reduction of 20% with respect to 2000 (Tierney, 2016).

[33] Coal production today is at its lowest level in at least three decades (Tierney, 2016).

[34] "In the US, the share of wind, solar, hydro, and other zero-carbon energy sources is expected to jump from 14% in 2015 to 44% in 2040" (Hood, 2016).

As indicated in the Annual Energy Outlook 2016 report, the US demand for coal might gradually decrease during the period 2014—40, instead of experiencing an immediate collapse in market share. The cumulative effects of market factors (flat demand in the electricity sector, competitive prices of natural gas in relation to coal, greater efficiency of gas-fired power plant, and the expansion of the number of energy renewable projects) indicate that coal is unlikely to recover its position before 2000, even without the implementation of the CPP by the US government. Trends already underway in the coal industry point to lower the overall demand for this type of energy source in the future, especially for electricity generation and heating, due to environmental considerations adopted by several countries (Tierney, 2016).

However, this trend may be altered due to the new conditions that currently exist within the coal industry, after the election of the new US president. President Trump wants to revive the coal industry in the US and relax regulations adopted by the former Obama administration that hurt its development prospects. The current US president began to fulfill his campaign promises to take a group of measures to reduce to the minimum the impact on the coal industry approved by the previous administration and abandoned the Paris Climate Agreement signed by the former President Obama.

It is important to highlight that on August 2018 the US administration has presented a proposal to leave the adoption of specific regulations on carbon dioxide emissions in the hands of the states, replacing the demanding environmental legislation of the former President Obama. The new standard aims to reduce energy costs incurred by companies. According to EPA interim administrator Andrew Wheeler, the new regulations will allow consumers to save money and workers to keep their jobs while protecting the environment. Compared to the CPP, these new regulations will allow reducing the energy costs of industries between 0,2% and 0,5% by 2025. Also, the new rules will ultimately reduce the CO_2 emissions expected for 2030 by around 1,5% thanks to an improvement in efficiency in industrial production. With these new measures, the US government will give the states a term of three years to establish their standards for the regulation of the use of coal, and the EPA must then approve that in one year. If the state proposal is not supported, then the federal government may implement its own regulations.

In summary, about the coal sector in the North America region, the following can be stated, according to the World Energy Resources 2016 report:

1. Coal is the second most important energy source worldwide. Hard coal and lignite (brown coal) are the primary sources of energy for electricity generation and heating, with 40% of the power generated worldwide.

2. Coal is predominantly a fossil fuel that is extracted and used in the same country, which guarantees its safety in its supply. The excess of supply and the price of natural gas, as well as the regulations regarding the protection of the environment and the population adopted by many countries have had a negative impact on the coal industry at world level, reducing the share of coal in the energy mix in many countries.

3. A total of 75% of all coal-fired power plants use subcritical technology. An increase in the operating efficiency of coal-fired power plants around the world from the current 33% to 40% could reduce the global emission of CO_2 by 1,7 billion tons per year.

4. In addition to the continuous increase in the efficiency of the operation of coal-fired power plants, the use and storage of carbon capture technology are one of the measures to be implemented for climate protection that could allow, in some countries, the use of this type of energy sources for electricity generation and heating at least until 2040.

5. Global coal consumption increased by 64% between 2000 and 2014, and for this reason, it is considered the fastest growing fossil fuel in absolute numbers within the indicated period. The years 2014 and 2015 were the first years since 1999 in which there was a decrease in global thermal coal production of 0,7% and 2,8%, respectively. It is likely that the use of coal as a fuel for electricity generation and heating will be reduced in several countries during the coming decades.

6. The use of new technologies that reduce the emissions of polluting gases into the atmosphere as a result of the operation of coal-fired power plants is essential to utilizing the abundant coal reserves in an increasingly carbon-constrained environment.

7. Oversupply and the price of natural gas have had a negative impact on the world coal industry (World Energy Resources 2016, used with the permission of the World Energy Council, www.worldenergy.org, 2016).

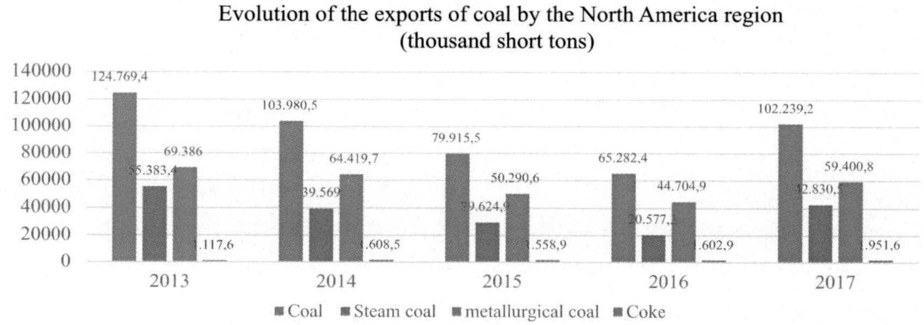

Figure 1.32 Evolution of the exports of coal by the North America region during the period 2013—17 (thousand short tons). *(Source: EIA database 2018.)*

Coal Exports from the North America Region

The US is a net exporter of coal. According to the American Geoscience Institute (2018), in 2017, the US exported 96.953.385 million short tons of coal to dozens of countries around the world; the most significant markets were India (15,4% of the total), South Korea (8,1% of the total), and Brazil (7,8% of the total). Out of this total, the US exports 41.698.819 million short tons of steam coal, 55.254.666 million short tons of metallurgical coal, and 1.209.947 million short tons of coke. In the case of Canada, the country exported, in 2017, a total of 5.285.938 million short tons of coal, 1.140.664 million short tons of steam coal, 4.145.274 million short tons of metallurgical coal, and 772.477 million short tons of coke.

In summary, the evolution of the exports of coal, steam coal, metallurgical coal, and coke by the North America region is shown in Fig. 1.32.

According to Fig. 1.32, the exports of coal from the North America region decreased during the period 2013-17 by 19,1% falling from 124.769,40 thousand short tons in 2013 to 102.239,30 thousand short tons in 2017. A similar situation can be found in the exports of steam coal with a decrease of 22,7% during the period considered, as well as in the exports of metallurgical coal with a decrease of 14,4%. The only type of coal that increased its exports from the region during the period considered is coke with 74,7%.

Coal Imports by the North America Region

In 2017, the total of coal imported by the North America region reached 8.668.906 short tons; a decrease of 20,7% concerning the level registered in 2016 (10.928.216 short tons).

Even though the US is a net exporter of coal, the country still imports some coal, mostly for its use in coal-fired plants for electricity generation and heating on the eastern and southern coasts of the US. For the US, it is cheaper to ship coal by sea from South America than transporting it from mines in the Northern and Western US (How Much Coal Does the US Exports and Import, 2018). In 2017, the US imports 7.777.172 short tons of coal, representing 89,7% of the total coal imported by the region in that year. Canada imports in the same year 891.734 short tons of coal or 10,3% of the total.

The evolution of the imports of coal, steam coal, metallurgical coal, and coke in the North America region during the period 2013—17 is shown in Fig. 1.33.

By the data included in Fig. 1.33, the following can be stated: the imports of coal by the North America region during the period 2013—17 decreased by 13,6%, falling from 10.022.501 short tons in 2013 to 8.668.906 short tons in 2017. It is predictable that the imports of coal by the region will continue to decrease during the coming years as a result of the measures adopted by the US government to reduce the use of coal for electricity generation and heating, and the closure of all coal-fired power plants approved by the Canadian government by 2030. In the case of the steam coal, the imports of this type of coal decreased by 10,9% during the period considered, falling from 8.108.312 short tons in 2013 to 7.229.377 short tons in 2017. The imports of metallurgical coal in the region decreased by 24,8%, dropping from 1.914.192 short tons in 2013 to 1.439.531 short tons in 2017. The decline in the imports of coke was of 63,4%.

Evolution of the imports of coal in the North America region during 2013-2017 (short tons)

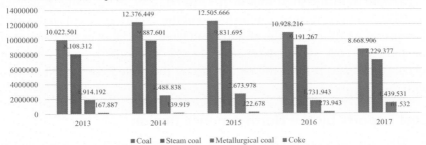

Figure 1.33 Evolution of the imports of coal, steam coal, metallurgical coal, and coke in the North America region during the period 2013—17 (short tons). *(Source: EIA database 2018.)*

Environmental and Social Impact for the Use of Conventional Energy Sources in the North America Region

Historically, the development and production of energy from fossil fuels have had a tremendous negative impact on the environment and human health. There is a great concern about the relationship between current methods of coal extraction, treatment of coal combustion waste, exploitation of natural gas, cleanup of oil spill pollution, and CO_2 emissions from the use of fossil fuels for electricity generation and heating, and the need to ensure stable and secure supplies of energy. The stable and reliable supply of coal should be done without damaging the environment or further altering the earth's climatic balance (Tarr, 2014).

On the other hand, the need to generate electricity most cleanly and safely has never been more evident to the international community. Without a doubt, environmental and health consequences of using oil and coal as a fuel for electricity generation and heating are essential issues, alongside the affordability of the power which is produced. The ecological and health implications of the use of fossil fuels for electricity generation and heating are generally considered as external costs; this means, those that are quantifiable but do not appear in the company's accounts. Therefore, they are not transmitted to the consumer but are borne by society in general. These external costs include, particularly, the effects of atmospheric pollution on human health, crop yields, as well as occupational diseases and accidents. Although they are even more difficult to quantify and evaluate than others, external costs also include effects on ecosystems and the impact of global warming.

The production of electricity from any form of primary energy source has not only some adverse environmental effect but also has some associated risks. A balanced evaluation of the different types of energy sources available in the country for electricity generation and heating requires, among other factors, a comparison of their environmental effects, including climatic changes and the impact on human health.

The environmental impacts of the use of fossil fuels for electricity generation and heating in the North American region can be summarized as follows:

- **Oil:** All energy production has a negative potential environmental impact, but not all of them on the same level. In the specific case of oil, it is a cleaner fossil fuel than coal, but not as clean as natural gas. Also, the use of oil for electricity generation and heating has certain disadvantages from the environmental point of view. The disadvantages include:

- Oil refining is a great polluter of the environment. The transformation of crude oil into petrochemical's products releases toxins into the atmosphere that are hazardous to human health and the ecosystem.
- The burn of crude oil for electricity generation and heating releases CO_2 and this contaminating gas increases global warning.
- Oil spill accidents cause considerable environmental damage when it occurs.
- Surface mining of tar sands or drilling in nature reserves produces ecological disturbance and a higher CO_2 emission than natural gas (Erbach, 2014).

- **Natural Gas:** Natural gas is one of the three fossil fuels that are currently used to generate electricity in many countries. Without a doubt, the emission level of polluting gases as a result of the use of natural gas for electricity generation and heating is much lower than those produced by the burn of coal or oil for the same purpose. According to some experts' calculations, natural gas emits between 50% and 60% less CO_2 when burned in a new and efficient natural gas-fired power plant compared to the emissions of gases in a typical coal-fired power plant (Cost and Performance Baseline for Fossil Energy Plants, National Energy Technology Laboratory, 2010). On the other hand, the electricity generation and heating produced by the burning of natural gas offers a variety of environmental and ecological benefits in comparison with the burning of coal and oil for the same purpose. These benefits are:
  - fewer emissions of less polluting gases into the atmosphere in contrast with the burning of other fossil fuel's sources, such as coal or oil;
  - natural gas is the preferred option for new cogeneration power plants as it increases the energy efficiency of electric generation systems and industrial boilers, and requires the burning of less fossil fuel to produce the same amount of electricity; and
  - natural gas combined cycle generation units can have up to 60% energy efficiency, while the energy efficiency of the coal and oil generation units is only between 30% and 35%.

The future climate impact of shale gas would be positive if it replaces dirty coal, and methane emissions can be minimized.[35] However, it is important to single out that it would be very harmful if cheap gas discourages investments in energy efficiency and renewable energy sources (Erbach, 2014).

In light of the considerable uncertainty regarding the size of the recoverable shale gas and tight oil resources, analysts are divided about the long-term prospects for energy production in the North America region. Some believe that the US is looking for a century of abundant energy supplies and even North America's energy independence, which according to their view will be reached in the coming years. This group of experts believes that the region will be transformed, in the future, as a net energy exporter. Others fear that the shale revolution will be a short-lived financial bubble, and predict energy scarcity and rising prices during the coming years. "Clearly, the way this happens will have a major impact on energy policies and the engagement of the US in other energy-producing regions, such as the Middle East" (Erbach, 2014).

- **Coal:** Coal-fired power plants emit more than 60 different hazardous air pollutants during electricity generation and heating. Despite billions of dollars of investment, with the aim of reducing the emission of contaminant gases for the burning of coal, scientists are unable to remove all harmful emissions from coal-fired power plants (The Environmental Impacts of Coal, 2005). According to McGuire (2001), "pollution from coal-fired power plants (Fig. 1.34) is released in four main ways:
  - as fly ash from the smoke stacks;
  - as bottom ash, which stays at the bottom after the coal is burned;
  - as waste gases from the scrubber units (which are chemical processes used to remove some pollutants); and
  - as gas released into the air."

---

[35] Currently, between 80% and 83% of global energy consumption is drawn from fossil fuels (Hood, 2016; BP Energy Outlook 2016, 2016). Fossil fuels are now and will remain the dominant sources of energy powering the world economy, supplying 60% of the energy increase up to 2035. In the specific case of the North America region, "fossil fuels remain the dominant forms of energy, accounting for 78% of total energy consumption in 2035, down from 83% today." Without a doubt, natural gas looks set to become the fastest growing fossil fuel supported by strong growth, particularly of US shale gas and LNG. The growth in LNG coincides with a significant shift in the regional pattern of trade. The US is likely to become a net exporter of gas later this decade, while the dependence of Europe and China on imported gas is projected to increase further (BP Energy Outlook 2016, 2016).

Figure 1.34 Big Ben coal power plant. *(Source: Wikimedia Commons.)*

The environmental impact on the use of coal as a fuel for electricity generation and heating can be summarized as follows:

- CO_2 emissions for the burning of coal exceed those of natural gas.
- "Coal-fired power plants produce large quantities of SO_2 and NO_X, the key pollutants in the formation of acid rain. Acid rain acidifies water bodies and harms forest and coastal ecosystems" (Keating, 2001). In the specific case of the NO_X, it helps in the formation of smog and nitrates, also affecting ozone (Effects of Acid Rain: Human Health, 2004).
- According to the Effects of Acid Rain: Human Health (2004) document, "coal-fired power plants are a major source of particulate pollution. Many scientific studies have shown that raised levels of particulates result in increased illness and premature death from heart and lung disorders, such as asthma and bronchitis" (Schneider, 2000).
- Coal contains numerous trace elements that are released during the burning of coal for electricity generation and heating. These elements are mercury,[36] dioxins, arsenic, radionucleotides, cadmium, and lead" (Keating, 2001). The production of energy and heat from coal-fired power plants is the largest source of mercury emissions into the atmosphere (Sources and Cycling of Mercury to the Global Environment, UNEP Global Mercury Assessment, 2002). According to the report mentioned above, "mercury and its compounds pose a global environmental threat to humans and wildlife." Mercury produces haze and can cause chronic bronchitis, aggravated asthma, and premature

---

[36] According to the Union of Concerned Scientists source, 42% of the US mercury emission is produced by coal-fired power plants.

death (both sulfur dioxide and nitrogen oxides transform into particulates in the atmosphere) (Coal Power: Air Pollution, 2008).

"Known human carcinogens and some of the most toxic compounds known to science, dioxins, can also be formed when coal is burned in fired-power plants for electricity generation and heating since most coal contains chlorine" (Coal-Fired Power Plants and the Menace of Mercury Emissions, 2001);

- Coal combustion waste (CCW), such as fly ash and bottom ash, and captured contaminants are usually disposed of in landfills or sold for industrial use. "Regardless of how the CCW is disposed of, there is a risk that the toxic metals are leaching into nearby surface and groundwater, contaminating said sources and rendering their consumption dangerous" (Keating, 2001).

Finally, it is important to emphasize that, to stop the negative impact on the environment and human health due to the use of certain fossil fuels for electricity generation and heating, such as coal and oil, all available renewable energy sources should increase their participation in the energy mix of the country in the coming years. For some countries, the use of nuclear energy could also be an excellent option to be considered. The problem is that the world should achieve this goal while global energy consumption continues to grow at an extremely ferocious rate in many countries, particularly in the most advanced developing countries. There is a need to find an effective way to use more and more clean energy available at world level for electricity generation and heating, since not only some fossil fuels such as oil, are being depleted, but this must be done to prevent the world from getting hot (Branson and Loewen, 2012).

Evolution of Energy Conventional Prices in the North America Region

For consumers, it is crucial that all energy sources can compete on equal terms, and provide the right signals to the market. In this sense, the granting of subsidies for the promotion of the use of specific energy sources plays an important role and must be carefully analyzed before their allocation. Recognizing the importance of applying a carbon emission penalty will undoubtedly be a key element in promoting adequate consumer behavior and will allow growth in investments in low-carbon power plants. That includes, among others, incentives to encourage investments in the use of renewable energies for electricity generation and heating, improvements in energy efficiency, and measures to guarantee the protection of the environment (World Energy Resources 2016, used with the permission of the World Energy Council, www.worldenergy.org, 2016).

According to the International Energy Outlook 2016 with Projection to 2040 report, expectations regarding future world oil prices are a source of uncertainty. Among the key factors that should be considered during the analysis of the evolution of world oil prices are:

- the investment and production decisions to be adopted in the future by OPEC member states;
- the supply of liquids from the non–OPEC countries; and
- the global demand for oil and other liquids in the future.

Also, the industry must make its predictions, at least in the general long-term movement of oil prices, since short-term changes tend to be driven by the news of the day. In particular, the industry must consider, according to Austin (2016), the following elements:

- supply and demand;
- political events;
- economic growth; and
- related markets.

Real oil prices, in US dollars of 2013, have fallen significantly since 2011, from approximately US$115 per barrel to around US$46 per barrel in 2017; that is, a reduction of 40%.[37] The production of 0,1 million barrels per day of combined oil production by Libya and Nigeria contributed to lower oil prices in 2017, as did the increase in inventories of crude oil and derivative products, which were above the two weeks average during the first two weeks of June.

EIA forecasts that world crude oil inventories will return to 0,2 million of barrels per day in 2018-2019. Given this expectation, Brent's crude oil prices are expected to remain reasonably stable over the coming months. However, some upward pressures could arise in the second half of 2019, if global crude oil inventories decrease during that period, and if the market expects a reduction of the comprehensive crude oil inventory by the end of 2019. EIA also forecasts that Brent's crude oil prices could rise to US$55 per barrel at the end of 2019 (Short-Term Energy Outlook, 2017).[38]

---

[37] In May 2018, the oil price increased up to US$77 per barrel.

[38] "Average West Texas Intermediate (WTI) crude oil prices were forecast to be US$2 per barrel lower than Brent's prices in 2017 and in 2018. The slight price discount of WTI to Brent in the forecast is based on the assumption that rising US crude oil production will result in WTI-priced US crude oil exports competing with international volumes priced off of Brent in global crude oil markets" (Short-Term Energy Outlook, 2017).

On the other hand, world crude oil prices are expected to recover during the period up to 2040, and it is forecast to reach US$141 per barrel in that year. It is estimated that the total world consumption of crude oil will grow 1,6% per year during the period 2012—40, well below 3,3% per year registered in the previous period, suggesting that a smaller portion of resources would be needed to cover consumers' crude oil needs. The International Energy Outlook 2016 with Projection to 2040 report reflects the expectations of the average cost that could be reached for exploration and development costs and access to crude oil resources, based on the hypothesis that OPEC producers will decide to maintain their participation in the global fuel liquids market between 39% and 43% during the coming years.

However, lower crude oil prices encourage consumers to use more liquid fuels, and this situation could have an impact on the level of crude oil prices until 2040. OECD nations are expected to consume 46,1 million barrels of crude oil per day by 2040. Higher crude oil prices, combined with the same level of economic activity, could cause oil consumers within the OECD to implement new energy efficiency measures and move to consume less expensive fuels where possible.

In summary, the following can be stated: the main elements that could cause the increase in crude oil price in the coming years are:

• a weaker dollar;
• level of crude oil inventories and the US storage space;
• level of crude oil production;
• the interruption of crude oil supplies; and
• the volatility in the crude oil market (Austin, 2016).

Compared to a year ago, it is expected that by 2040, natural gas and coal prices will be significantly lower than current natural gas and coal prices. "In the US, natural gas at competitive prices, rapid growth in the use of renewable energies for electricity generation and heating, and the adoption of strict regulatory measures on the use of coal in power plants will limit the use of coal for electricity generation" (BP Energy Outlook Focus on North America, 2016).

Electricity Generation and Consumption in the North America Region

There is no doubt, that "the worldwide mix of primary fuels used to generate electricity has changed a great deal over the past several decades" (International Energy Outlook 2016 with Projection to 2040, 2016).

For example, electricity generation from natural gas-fired power plants increased considerably after the 1980s. On the other hand, the use of petroleum and other liquids for the same purpose declined after the late 1970s, when sharp increases in oil prices encouraged power generators to substitute oil for other energy sources (International Energy Outlook 2016 with Projection to 2040, 2016). Despite those changes, coal was and will continue to be the fossil fuel most widely used in electricity generation and heating in many countries, and this situation will not change at least until 2040, according to the report mentioned above.

Fig. 1.35 shows the foreseeable evolution of global net electricity generation during the period 2012—40.

According to Fig. 1.35, the global net electricity generation is expected to increase by 69% during the period 2012—40, rising from 21,6 trillion kWh in 2012 to 36,5 trillion kWh in 2040. This increase is due to a growth in the demand for electricity in the most advanced developing countries located in different regions. Undoubtedly, power is, and will continue to be, the fastest-growing final form of energy consumption throughout the world during the coming years, as it has been for many decades, and this trend is expected to continue in the future, transforming isolated and noncompetitive network systems into integrated national and even regional and international markets (International Energy Outlook 2016 with Projection to 2040, 2016).

The most substantial growth in electricity generation and heating is expected to occur mainly among developing countries with an expected average increase of 2,5% per year during the period 2012-40. This projected increase is the result of an increase in the standard of living of the population of these countries, and an increase in electricity demand due to the enlarged use of electrical appliances and other electronic devices, as well

Expected evolution of the global net electricity generation
during the period 2012-2040 (trillion kWh)

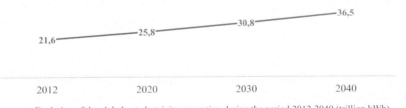

— —Evolution of the global net electricity generation during the period 2012-2040 (trillion kWh)

Figure 1.35 Expected evolution of global net electricity generation during the period 2012—40 (trillion kWh). *(Source: International Energy Outlook 2016 with Projection to 2040, 2016. US Energy Information Administration, Office of Energy Analysis, US Department of Energy, IEO 2016.)*

as to meet the energy demand for commercial services, hospitals, schools, office buildings, among others.

In most developed countries, where infrastructures in the energy sector are more mature than in developing countries, several measures to increase energy efficiency in the electricity generation sector have been adopted, or are expected to be approved shortly. It is important to highlight that where the population growth is relatively slow or decreasing, energy generation is expected to increase by an average of 1,2% per year during the period 2012–40, half of the projected average increase for developing countries (2,5%).

The expected evolution in the participation of all energy sources in the world electricity generation during the period 2012–40 is shown in Fig. 1.35. According to that figure, natural gas, renewables, and coal are the primary energy sources more used at world level for electricity generation and heating in 2040 (Fig. 1.36).

In the specific case of the North America region, the evolution of the electricity generated in the area during the period 2010–17 is shown in Fig. 1.37.

Looking at Fig. 1.37, it can be stated that the total electricity generation in the North America region decreased by 0,6% during the period 2010–17, falling from 5.001,2 TWh in 2010 to 4.975,2 TWh in 2017. The peak in the production of electricity during the period under consideration was reached in 2014 and the minimum level in 2012. It is projected that the electricity generation in the region will continue to move in the same manner; some years will increase while in others it will decrease, depending on the measures adopted by the US and Canada governments to increase the efficiency of the electrical equipment use by the population and the industry.

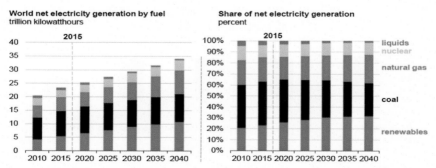

Figure 1.36 World net electricity generation by energy sources during the period 2012–40 (trillion kWh). *(Source: International Energy Outlook 2017, September 2017. Energy Information Administration (EIA).)*

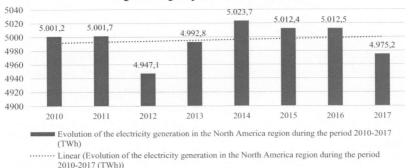

Figure 1.37 Evolution of the electricity generation in the North America region during the period 2010–17. *(Source: BP Statistical Review of World Energy 2017, June 2017.)*

The evolution of the electricity generation using fossil fuels as fuel in the North America region during the period 2012–17 is shown in Fig. 1.38.

According to Fig. 1.38, the electricity generation in the North America region during the period 2012–15 using fossil fuels as fuel decreased 7,4% from 2.903 billion kWh in 2012 to 2.689 billion kWh in 2017. Because the US and Canada are reducing the use of coal and petroleum and other liquids for electricity production, it is probable that the role of fossil fuels as part of the energy mix of these countries will continue to fall during the coming years. At the same time, the use of natural gas for electricity generation and heating is expected to be the only fossil fuel to increase its participation in the energy mix of these two countries during the coming years.

On the other hand, it is projected that up to 2040 the share of petroleum and other liquids in the global electricity generation will decrease 45,5%,

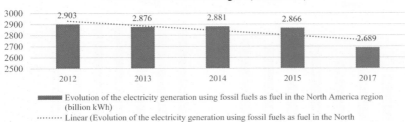

Figure 1.38 Evolution of the electricity generation using fossil fuels as fuel in the North America region during the period 2012–17. *(Source: EIA database 2018 and author's own calculation for 2017.)*

falling from 1,1 TWh in 2012 to 0,6 TWh in 2040 (International Energy Outlook 2016 with Projection to 2040, 2016). According to the report mentioned before, the percentage in the participation of liquid fuels in the world's electricity generation is projected to decrease from 5% in 2012 to less than 1,6% in 2040; that is a reduction of 3,4%. It is expected that this trend will continue after 2040.

In the North America region, the participation of liquids in the energy mix of the US and Canada is the lowest of the three types of fossil fuels used for electricity generation and heating, and it is foreseen that this situation will not change during the coming years. In the specific case of natural gas, the use of this type of energy source for electricity generation and heating is expected to grow 6%, rising from 22% in 2012 to 28% in 2040. It is likely that after renewable energy sources, natural gas will be the fastest second source of the world's electricity generation up to 2040.

Many countries have adopted specific policies and enacted environmental regulations aimed at reducing the consumption of fossil fuels, especially coal and oil, by power plants during electricity generation and heating. The aim is to decrease, as much as possible, greenhouse gas emissions during the operation of this type of power plants. As a result of the efforts made by most of the countries, it is expected that the participation of coal as a fuel for the operation of power plants will decrease considerably during the coming decades. At the end of the period 2012—40, the electricity production by renewable energy sources is expected to be identical to the electricity generation from power plants using coal as fuel throughout the world.

According to the International Energy Outlook 2016 with Projection to 2040 report, the use of coal for electricity generation and heating in 2012 represented around 43% of the world consumption of this type of energy source. In 2020, it is likely that the share of coal in world energy consumption will be 40%, and it is projected that this percentage will be lower in 2040 (around 31%) (see Fig. 1.39).

Electricity Trade Within the North America Region

It is important to highlight that Canada and the US have been commercial partners in the exports and imports of electricity for more than a century.[39] Undoubtedly, a stable flow of power is a crucial element in the

[39] According to the Energy Fact Book 2016—17, in 2015, all Canadian electricity trade was with the US representing 9% of the total electricity generated by the country in that year, and is used to meet 2% of the US electricity consumption.

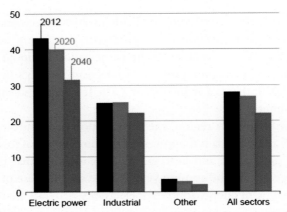

Figure 1.39 Coal share of world energy consumption by sector during the period 2012–40. *(Source: International Energy Outlook 2016 with Projection to 2040, 2016. US Energy Information Administration, Office of Energy Analysis, US Department of Energy, IEO 2016.)*

development of electricity trade beneficial to both countries. The existence of multiple connections makes the entire North American electricity grid more stable than many others around the world.

Canada's net electricity exports reached record export levels of 73,1 TWh in 2016. The export sales in 2016 and 2017 reached 73,1 TWh and 72,1 TWh, respectively. Canada's income for electricity exports reached US$2.900 million in 2017, similar to the level reached in 2016. The highest export sales were achieved in 2015 (US$3.100). The values of Canada's sales in the export of electricity increased by more than 17% in 2015 in comparison to the levels reached in 2014 and decreased 1,4% during the period 2016–17 (Electricity Trade Summary National Energy Board, March 2018).

As happened in these years, the provinces with large installed hydrological capacities, Quebec, Ontario, Manitoba, and British Columbia, obtained almost 95% of the total value of the reported export sales of electricity. While record on export sales of power has not yet been translated into record export earnings, revenues from this concept continue to rise, and it is expected that this trend will continue without change during the coming years.

One crucial element that needs to be considered is the fact that the Canadian annual export sales of electricity to the US varied significantly (see Fig. 1.40). In the last years, electricity export sales averaged almost 60 TWh, ranging from 51,2 TWh in 2011 to 72,1 TWh in 2017.

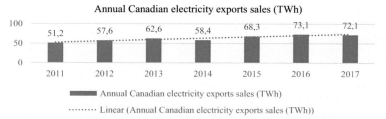

Figure 1.40 Annual Canadian electricity export sales during the period 2011–17. *(Source: Electricity Trade Summary of the National Energy Board of Canada, 2018.)*

Historically, the years of high amounts of electricity exports coincide, generally, with years of high precipitation, and the provinces with hydro-electric generation exported the most substantial amounts of energy. The peak in the electricity export sales by Canada was reached in 2016.

According to Fig. 1.40, Canadian electricity export sales increased 40,8% during the period 2011–17, with a record high registered in 2016. It is projected that Canadian electricity export sales will continue to grow during the coming years, and the US will continue to be the leading partner.

Overall, Canada is a net exporter of electricity to the US. Most of the US power needs are met by hydroelectricity and not by fossil fuels. On a net basis, Canada electricity exports went mainly to New England, New York, and the Midwest states, while the US electricity exports went primarily from the Pacific Northwest states to the Canadian provinces of British Columbia, Manitoba, and Quebec (Manzagol and Hodge, 2015). The value of the electricity exports by Canada, in 2017, reached US$2,1 billion, representing 7,7% of the world total. Based on this exports value, Canada is ranking third in the list of main electricity exporter countries in value terms.

In general, the North America region exported electricity, in 2017, equivalent to 8,9% of the total after Europe with 66,6% and Asia with 12,3%.

Canadian electricity import purchases were at their lowest in almost 20 years, decreasing by more than 30% from 2014 to 2015. After 2015, the electricity import purchases by Canada increased 13,8% from 2015 to 2017. Imports are about 25% below the five year average of 11,5 TWh. In 2017, electricity imported by Canada was 9,9 TWh (see Fig. 1.41).

According to Fig. 1.41, the electricity imports by Canada decreased 31,3% during the period 2011–17, falling from 14,4 TWh in 2011 to 9,9 TWh in 2017. The peak in the electricity imports by Canada within the period considered was reached in 2011. During the period 2015–17,

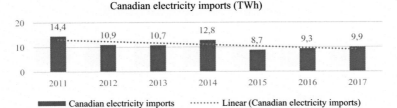

Figure 1.41 Annual Canadian electricity import purchases during the period 2011—17. *(Source: International Trade Statistics (March 2016) and Electricity Trade Summary of the National Energy Board of Canada, 2018.)*

the electricity imports by Canada increased by 13,8%. It is expected that the electricity import purchases will continue to grow during at least the coming years, but perhaps at a lower rate.

In the case of the US, the electricity exports during the period 2012—16 decreased 48,4%, reached its lowest level in 2016 (see Fig. 1.41). The peak in electricity exports was achieved in 2014. Most of the electricity exports from the US during the period considered went to Mexico and Canada.

The evolution of the electricity exports from the US during the period 2012—16 is shown in Fig. 1.42.

According to Fig. 1.42, the following can be stated: the US electricity exports during the period 2012—16 decreased by 48,4%, falling from 12 billion kWh in 2012 to 6,2 billion kWh in 2016. The peak in the US electricity exports within the period considered was reached in 2014. Since that year the exports of power by the US decreased significantly (52,4%). By the data included in the above figure, it can be foreseen that the electricity exports from the US, particularly to Canada, will continue to decrease at least during the coming years because Canada's import needs are expected to continue to be very low during the same period.

The evolution of the electricity imports by the US during the period 2012—16 is shown in Fig. 1.43.

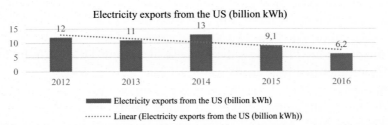

Figure 1.42 Evolution of the electricity exports from the US during the period 2012—16. *(Source: EIA database 2017.)*

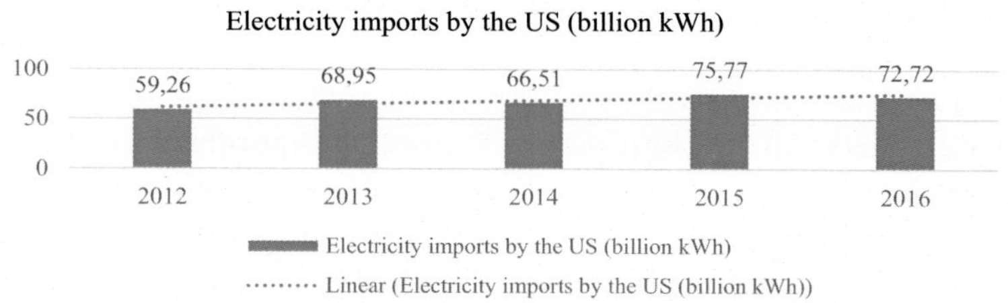

Electricity imports by the US (billion kWh)

Figure 1.43 Evolution of the electricity imports by the US during the period 2012—16 (billion kWh). *(Source: EIA database 2017.)*

Based on the data included in Fig. 1.43, the following can be stated: the imports of electricity by the US increased by 22,7% during the period 2012—16, rising from 59,26 billion kWh in 2012 to 72,72 billion kWh in 2016. The peak in the electricity imported by the US was reached in 2015. Most of the electricity imports by the US are from Canada. It is anticipated that this situation will continue without change during the coming years.

There should be no doubt that, in the long-term, the US electricity trade with Canada is expected to increase, providing more economical and reliability benefits to both states. Canada will continue to be a net exporter of electricity to the US, while the US will continue to be a net importer of electricity from Canada (Canada Energy Future 2013. Energy Supply and Demand Projection to 2035, 2013). Although the amount of electricity imported over the Canadian border is a small part of the overall US power supply, the transmission connections linking Canada and the US facilitate the electricity trade among them and are an essential component of the electricity markets in northern states (Manzagol and Hodge, 2015).

In summary, the following can be stated: according to the report of the Canada National Energy Board (2013), the electricity generation in the North America region has changed over recent years, with implications for the consumption of coal, natural gas, and renewable energy. While the share of coal-fired power plants used for electricity generation and heating has decreased and will continue this trend during the coming years, the percentage of natural gas and non–hydro renewable energy has increased in response to changes in economic conditions, government regulations, and environmental concerns.

Despite all measures adopted by the US and Canada to reduce the share of coal-fired power plants in the electricity generation and heating in both countries, coal remains, in 2015, the most common source of electricity in the US, but followed very closely by natural gas. The decrease in the use of coal as an energy source for electricity generation and heating in both countries is due, in large part, to the increased use of natural gas and the adoption of policy initiatives, such as the elimination of the use of coal for electricity generation and heating in Ontario, among others. In the short-term, the shift away from the use of coal-fired power plants for electricity generation and heating to the use of natural gas-fired power plants with this same objective occurs when the price of natural gas is competitive with the price of coal. Recently, the electricity generation due to the use of natural gas-fired power plants equaled or exceeded the electricity generation from coal-fired power plants. In April 2012, coal and natural gas accounted for 32% of total electricity generation in the US. In the same year, in Canada, the share of electricity generation from natural gas-fired plants increased above electricity production from coal-fired power plants.

Without a doubt, the US and Canadian government energy policies, particularly in the case of Canada, discourage the use of coal for electricity generation and heating. Canadian federal regulations enacted in 2012 stipulate that coal-fired power plants that start operating after July 1, 2015 must not emit more than 420 metric tons of CO_2 per GWh, often known as a level consistent with the high-efficiency level reached by natural gas-fired power plants. Existing coal-fired power plants that do not meet this standard will be closed after 50 years of operation, or at the end of 2019 or 2029 (depending on their start-up date), whichever time comes first. Coal-fired power plants will probably have to install carbon capture and storage equipment to comply with these new rules. As a result, it is likely that the share of electricity generation based on coal-fired power plants will decrease steadily in Canada until 2030 when all coal-fired power plants must be closed (Canada Energy Future 2013. Energy Supply and Demand Projections to 2035, 2013).

In the US, several regulations now impose a limit on the level of mercury, sulfur dioxide, nitrogen oxides, and other pollutants emitted by coal-fired power plants. At the same time, lower capital costs provide an incentive for the construction of natural gas-fired power plants instead of power plants using coal as a fuel for electricity generation and heating in the country.

The investments needed to ensure compliance by coal-fired power plant operators with the new standards adopted by the US and Canadian governments will increase the cost of generating electricity by coal-fired power plants in both countries. According to the EIA projection, only 3% of the energy capacity additions in the US to be built up 2040 will come from coal-fired power plants. In contrast, the EIA foresees that natural

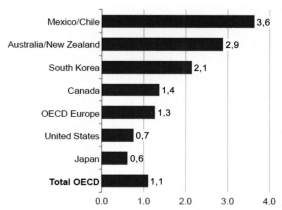

Figure 1.44 Average annual change in OECD commercial sector consumption during the period 2012–40 (percentage per year). *(Source: International Energy Outlook 2016 with Projection to 2040, 2016. US Energy Information Administration, Office of Energy Analysis, US Department of Energy, IEO 2016.)*

gas-fired power plants will provide 63% of additional electricity generation to be built in that period and renewable energy the remaining 31%.

Electricity Generation in the Commercial Sector

The US has the highest energy use in the world's commercial sector, as a result of its high average per capita income (measured as the average GDP per person), which is also among the highest in the world. A large part of the economic activity of the US is concentrated in the commercial sector. Among the OECD countries, the US is projected to have the second-slowest growth rate in the consumption of energy during the period 2012–40, after Japan (see Fig. 1.44). However, in recent years the increase in the use of US commercial energy has been relatively slow as a result, partially, of the adoption of "federal efficiency standards,[40] which foster technological improvements in end-use equipment and act to limit growth in delivered energy consumption relative to growth in commercial floor space" (Annual Energy Outlook, 2013).

In the case of Canada, the consumption of natural gas in the commercial sector in 2014 reached 1,33 billion cubic feet or 19,1% of the total. The annual average change in the commercial sector energy consumption

---

[40] Efficiency improvements in the US are likely for lighting, refrigeration, space cooling, and space heating, as a result of the implementation of the Energy Independence and Security Act of 2007 and the Energy Policy Act of 2005.

for the period 2012—40 is expected to be 1,4%, double than that projected for the US.

The total electricity consumed by Canada in the commercial sector in 2014 represented 21,5% of the entire power generated by the country in that year (Electricity Facts, Natural Resources Canada, 2018).

Electricity Generation in the Residential Sector

Residential electricity consumption in the US and Canada represented over a quarter of total residential electricity consumption in 2012. The US is the largest consumer of residential electricity within the OECD Americas, accounting for 84% of the total in 2012. However, this percentage is expected to be reduced to 80% in 2040; that means a reduction of 4% concerning the portion registered in 2012 (0,14% annual decrease on an average). In other words, residential electricity consumption in the US is expected to remain almost constant between 2012 and 2040, "as state and federal energy efficiency standards for consumer products, other efficiency programs, and sectoral changes offset drivers who would otherwise increase residential energy demand" (International Energy Outlook 2016 with Projection to 2040, 2016).

In the case of Canada, "the higher penetration of heat pumps for heating and cooling in buildings leads to lower energy use in the residential sector because that technology heats and cools more efficiently than conventional natural gas, electric, or oil systems" (Canada's Energy Future 2017. Energy Supply and Demand Projections to 2040, 2017). The residential sector consumed, in 2014, a total of 1,82 billion cubic feet per day of natural gas or 26,2% of the total (Natural Gas Facts, Natural Resources Canada, 2018). The total electricity consumed by the commercial sector in Canada reached 33,1% of the whole power generated by the country in 2014 (Electricity Facts, Natural Resources Canada, 2018). It is likely that the average annual growth in the use of electricity in the residential sector will be 0,4% during the period 2016—40.

The Contribution of the Different Conventional Energy Sources to Electricity Generation in the North America Region

The contribution of the different conventional energy sources to the electricity generation in the US is shown in Table 1.14.

Table 1.14 Evolution of the Electricity Generation by Different Conventional Energy Sources in the US During the Period 2010—15 (thousand MWh).

Period	Coal	Petroleum Liquids	Petroleum Coke	Natural Gas	Other Gas
2010	1.847.290	23.337	13.724	987.697	11.313
2011	1.733.430	16.086	14.096	1.013.689	11.566
2012	1.514.043	13.403	9.787	1.225.894	11.898
2013	1.581.115	13.820	13.344	1.124.836	12.853
2014	1.581.710	18.276	11.955	1.126.609	12.022
2015	1.352.398	17.372	10.877	1.333.482	13.117

Source: EIA database.

According to Table 1.14, the participation of coal, petroleum liquids, and petroleum coke in the energy mix of the US decreased 26,8%, 25,6%, and 20,8%, respectively, during the period 2010–15. It is expected that the level of participation of coal, petroleum liquids, and coke for electricity generation will be lower during the coming years. At the same time, the involvement of natural gas and other gas in the energy mix of the US during the period considered increased 35% and 15,9%, respectively. It is predictable that the participation of natural gas in the energy mix of the US will continue rising during the coming years.

In the case of Canada, in 2015, according to Statista database, the installed electricity generation capacity reached 140,7 GW. In 2016, the capacity installed in the country increased by 1,9% with respect to the year before, reaching 143,4 GW. The evolution of the participation of oil, natural gas, and coal in electricity generation in Canada during the period 2010–16 is shown in Fig. 1.44. It is important to highlight that fossil fuels are the second most important source of electricity in Canada. About 9,5% of electricity supply comes from coal, 8,5% from natural gas, and 1,3% from petroleum. Fossil fuel generation is particularly vital in Alberta and Saskatchewan, where several power plants have been built adjacent to large coal deposits. Fossil fuel generation is also significant in the Atlantic Provinces, Northwest Territories, and Nunavut. Ontario used to rely heavily on coal-fired generation but, in April 2014, the last coal-fired generating capacity was shut down (About Electricity. Natural Resources Canada, 2016).

In 2015, Canadian electricity demand was 522 TWh and accounted for 17% of total Canadian end-use energy demand. During the period 1990–2015, Canadian electricity demand increased by an average of 1% per year.

It is important to note that the majority of additions to the generation capacity in Canada during the period 2016–40 are expected to be in power plants that use natural gas as fuel, as well as in wind farm and hydro facilities, accounting for 85% of the new 54 GW capacity to be built in that period. The remaining additions include 0,4 GW of coal equipped with CCS technology. Coal, nuclear, and oil-fired power plants, in addition to some older natural gas-fired facilities, will make up the bulk of retirements to be carried out during that period.

According to Fig. 1.45, the use of natural gas for electricity generation and heating in Canada during the period 2010–17 increased by 52,9%, rising from 48 TWh in 2010 to 73,4 TWh in 2017. In that year, natural gas generated 10,6% of the total electricity produced in the country. In 2040, it

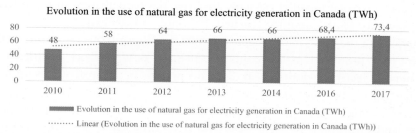

Figure 1.45 Evolution of the participation of natural gas in electricity generation and heating in Canada during the period 2010–17. *(Source: Canada's Energy Future 2017. Energy Supply and Demand Projections to 2040, 2017. National Energy Board; An Energy Market Assessment and BP Statistical Review of World Energy, June 2018.)*

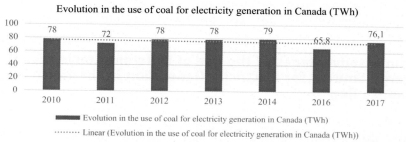

Figure 1.46 Evolution of the participation of coal in electricity generation and heating in Canada during the period 2010–17. *(Source: Canada's Energy Future 2017. Energy Supply and Demand Projections to 2040, 2017. National Energy Market Assessment and BP Statistical Review of World Energy, June 2018.)*

is projected that natural gas will generate 128 TWh or 17,5% of the total power produced in the country. Based on this previous information the following can be stated: during the period 2010–40, the use of natural gas for electricity generation and heating in Canada will increase 2,7 times, rising from 48 TWh in 2010 to 128 TWh in 2040.

Finally, the participation of coal in electricity generation and heating in Canada during the period 2010–17 is shown in Fig. 1.46.

By the information included in Fig. 1.46, the following can be stated: the use of coal for electricity generation and heating in Canada during the period 2010–17 decreased by 2,5% falling from 78 TWh in 2010 to 76,1 TWh in 2017. In this last year, coal generated 10,6% of the total electricity produced in the country.

On the other hand, and in the case of oil-fired power plants, the use of oil for electricity generation and heating in Canada accounted, in 2014, for

2,5% of the total installed capacity. In 2016, the participation of oil in the energy mix of the country reached 1% only and generated 26 TWh or 0,0059% of the total; in 2017 generated 22,7 TWh or 0,0047% of the total electricity produced in the country. Oil-fired power plants are used to generate electricity during peak demand periods or in areas where other generation options are not widely available.

According to the Canada's Energy Future 2016, Energy Supply and Demand Projections to 2040 report, total oil-fired capacity is expected to decline from 3,4 GW in 2014 to 2,7 GW in 2040; this means a decrease of 20,6% for the whole period. This reduction in the use of oil for electricity generation and heating reflects the retirements of aging units. In 2017, oil-fired power plants electricity generation currently accounts for roughly 0,5% of the total electricity produced in the country. It is projected that this percentage will continue to be at this low level over the projection period.

Main Disadvantages of the Use of Different Conventional Energy Sources for Electricity Generation in the North America Region

It is impossible to use the different conventional energy sources for electricity generation and heating available in the North America region without considering their disadvantages for the population and the environment. The main problems of the use of each type of conventional energy source for electricity generation and heating in the North America region can be summarized as follows:

Crude Oil

- The consumption of crude oil for electricity generation and heating produces carbon dioxide, among other contaminating gases, which causes greenhouse effect.
- The use of crude oil for electricity production and heating requires a significant and reliable supply of cooling water.
- The world's crude oil reserves are limited and, for this reason, cannot be considered the only long–term solution to energy needs by any country, even for those with the higher reserves of this type of energy source.
- If the crude oil spills while transporting, it causes severe pollution and affects the marine species, with considerable damage to the population and the environment.
- Some crude oils contain high levels of sulfur.

Natural Gas

- Natural gas is the less contaminated fossil fuel that can be used for electricity generation and heating. However, it is important to stress that when natural gas is burned, several contaminated gases are emitted into the atmosphere contributing to the greenhouse effect.
- The use of natural gas for electricity generation and heating is cheap in comparison with the use of oil and coal for the same purpose.
- Natural gas transportation cost is high in contrast to the transportation costs of other fossil fuels.
- During the transportation or in any another situation, natural gas is hazardous. Leaks can cause fire or explosions. The gas itself is extremely toxic when inhaled (Venkata Lalitha et al., 2014)).
- Natural gas reserves are limited and, for this reason, cannot be considered the only long-term solution to the energy supply problem by any country.
- The whole natural gas pipeline installation is costly to construct since long pipelines, specialized tanks, and separate plumbing systems need to be used. Pipeline leakage may also be very expensive to detect and repair (Aggeliki, 2014).

Coal

- Coal reserves are limited and, for this reason, cannot be considered the only long-term solution to energy needs by any country.
- The operation of a coal-fired power plant requires large amounts of coal to produce less amount of electricity in comparison with oil and natural gas-fired power plants, causing wastage of fuel (Venkata Lalitha et al., 2014).
- Coal is the most contaminating fossil fuel that can be used for electricity production and heating. For this reason, several countries are adopting measures to eliminate or drastically reduce the use of coal for electricity generation and heating during the coming years.
- The consumption of coal for electricity generation and heating requires more water than the use of oil and natural gas for the same purpose.
- The consumption of coal for electricity generation and heating produces large quantities of ash that have to be disposed of, and a lot of smoke (Venkata Lalitha et al., 2014).
- The methods of mining coal can be very destructive to the environment affecting the whole territory where it is located (Venkata Lalitha et al., 2014).

Summing up the following can be said, according to the BP Statistical Review of World Energy 2018 report:

- Global primary energy consumption grew strongly in 2017, led by natural gas. Coal's share of the energy mix continue to decline.
- Primary energy consumption growth averaged 2,2% in 2017, up from 1,2% in 2016 and the fastest since 2013.
- By fuel, natural gas accounted for the most substantial increment in energy consumption, followed by crude oil.
- The oil price (Dated Brent) averaged US$54,19 per barrel in 2017, up from US$43,73/barrel in 2016, the first annual increase since 2012.
- Global oil consumption growth averaged 1,8%, or 1,7 million barrels per day, above its 10-year average of 1,2% for the third consecutive year;
- Refinery throughput rose by an above-average 1,6 million barrels per day, while the refining capacity growth was only 0,6 million barrels per day, below average for the third consecutive year. As a result, refinery utilization climbed to its highest level in nine years.
- Natural gas consumption rose by 3.390,21 billion cubic feet (96 billion cubic meters), or 3%, the fastest since 2010.
- Consumption growth was driven by China (1.094,75 billion cubic feet or 31 billion cubic meters), the Middle East (988,81 billion cubic feet or 28 billion cubic meters), and Europe (918,18 billion cubic feet or 26 billion cubic meters). Consumption in the US fell by 1,2%, or 388,46 billion cubic feet.
- Global natural gas production increased by 4.626,22 billion cubic feet, or 4%, almost double the 10-year average growth rate.
- Natural gas trade expanded by 2.224,82 billion cubic feet, or 6,2%, with growth in LNG outpacing in pipeline trade.
- The increase in natural gas exports was mainly driven by Australian and US LNG (up by 600,35 and 459,09 billion cubic feet, respectively).
- Coal consumption increased by 25 million tons of oil equivalent, or 1%, the first growth since 2013.
- Coal's share in primary energy fell to 27,6%, the lowest since 2004.
- World coal production grew by 105 million tons of oil equivalent or 3,2%, the fastest rate of growth since 2011. Production rose by 56 million tons of oil equivalent in China, and 23 million tons of oil equivalent in the US.

- Power generation rose by 2,8%, close to the 10–year average. Practically all growth came from emerging economies (94%). Production in the OECD has remained relatively flat since 2010.
- Renewables accounted for almost half of the increase in power generation (49%), with most of the remainder provided by coal (44%).

CHAPTER 2

Current Status and Perspective in the Use of Crude Oil for Electricity Generation in the North America Region

Contents

Conventional Energy in North America
ISBN 978-0-12-814889-1
https://doi.org/10.1016/B978-0-12-814889-1.00002-4

General Overview

Crude oil is the product of the transformation of old organic material subjected to compression and heating for a long time. Once extracted, crude oil is refined in specialized facilities (refineries) built for that purpose, and in this way creates multiple types of products widely used by humans in their activities such as gasoline, diesel, and heating oil, among others. Due to its high-energy density, easy transportability, and relative abundance, crude oil has been the primary source of energy in the world since the mid-1950s. Like the contribution of coal to the Industrial Revolution, the transformation of crude oil into different useful products for human beings brought incredible advances in the economic development of different countries, particularly in the sector of transport of personnel and in other non-electrical applications, first in America and then in other parts around the world (Oil-IER, 2017).

According to the World Energy Outlook 2016 report, natural gas and crude oil are, and will continue to be, cornerstones of the regional and global energy system over the coming years. The increased use of natural gas will occur as a result of the commitments assumed by governments regarding the reduction of the adverse effects of human activity on the environment and climate change, and the decrease in the use of certain fossil fuels for electricity generation and heating in several countries.

However, the use of crude oil for electricity generation and non-electricity purpose, and its negative impact on the environment and population cannot be ignored by the industry. Several experts have said that the industry is not in a position to ignore the risks that could arise from a faster and deeper substitution of certain fossil fuels, such as coal and crude oil, as fuel for electricity generation and heating, by other energy sources such as natural gas and renewables. It has been estimated that, by 2040, the demand for crude oil will return to the levels reached in the late 1990s; this means a market with less than 75 million barrels per day.[1] The use of coal is expected to return to levels seen for the last time in the mid-1980s, with less than 3.000 million tons of carbon equivalent per year. Only natural gas is

---

[1] Other sources such as the International Energy Outlook 2017, estimate that for 2040 world consumption of crude oil and other liquid fuels will rise from 95 million barrels per day in 2015 to 113 million barrels per day in 2040; this represents an increase of 18,9%.

projected to register an increase in its use as fuel for electricity generation and heating worldwide.

Undoubtedly, crude oil remains the primary fuel in use worldwide, accounting for 32,9% of global energy consumption, mainly in the transport sector (63% of crude oil consumption). In 2040, the share of crude oil in the world energy consumption will decline to 30%, according to the International Energy Outlook 2016 with Projection to 2040 report. While emerging economies continue to dominate the growth of global energy consumption, the increase in these countries (+1,6%) was well below its average of the last 10 years (+3,8%). "Emerging economies now account for 58,1% of the global energy consumption" (BP Statistical Review of World Energy 2016, 2016). Chinese energy consumption growth slowed to only 1,5%, while India recorded another sharp increase in energy consumption (+5,2%). Energy consumption in OECD countries increased slightly (+0,1%), compared to an average annual decrease of 0,3% in the last decade. In 2015, a rare increase in the consumption by the countries of the European Union (EU) of 1,6% offsets the decrease recorded in the US of 0,9% and in Japan of 1,2%, where consumption fell to the lowest level since 1991 (IEA, 2016).

The evolution in the world consumption of crude oil during the period 1990−2040 is shown in Fig. 2.1.

According to Fig. 2.1, most of the growth in world crude oil and other liquids consumption during the period 2015−40 is expected to come from

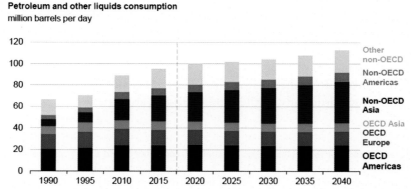

Figure 2.1 Evolution in the world consumption of crude oil during the period 1990−2040 (million barrels per day). *(Source: International Energy Outlook 2017 (2017).)*

non–OECD countries,[2] where strong economic and population growth foresee an increase in the demand for crude oil and other liquid fuels by 39%. In OECD countries, the need for crude oil and other liquid fuels is projected to grow slowly or decline between 2015 and 2040. Across sectors, the world share of crude oil and other liquids use is projected to be relatively constant during the period considered. The expected change is the following:

- In the transportation sector, liquids consumption is likely to increase by only 2%, rising from 54% in 2015 to 56% in 2040. This sector is and will remain the largest consumer of refined oil and other liquids.
- In the industrial sector, it is expected that no change will be registered within the period considered keeping its level of participation at 36%.
- In the building sector, a decrease is projected in the use of liquefied petroleum gas to provide a space heat of 1%, falling from 6% in 2015 to 5% in 2040.
- In the electricity sector, the use of liquid fuels is expected to decrease by 2% dropping from 4% in 2015 to 2% in 2040.

Several factors are affecting the use of liquid fuels to generate electricity. The factors include an increase in crude oil prices and relatively less costly natural gas, which are encouraging producers to reduce the use of crude oil and increase the use of natural gas for electricity generation and heating.

It is evident that crude oil is one of the essential commodities in the world. This characteristic of this type of energy source and the central role that crude oil and oil products play in modern life around the globe highlight its importance for the economy and the politics of the oil industry in the international arena.

Several structural changes have been taking place in the oil industry during the last years. Some of the main changes that have been registered are:

- the appearances of new non–OPEC oil supplying countries;
- current trends in the increase of energy efficiency recorded in many countries, particularly in the most advanced countries;

---

[2] More than 80% of the total increase in crude oil and other liquid fuels consumption is expected to be registered in non–OECD Asia, particularly in China and India, as a result of their rapid industrial growth and increased demand for transportation. China's use of crude oil and other liquid fuels for transportation is projected to increase by 36% during the period 2015–40 and India's use over that period is expected to increase by 142% (International Energy Outlook, 2017, 2017).

- the diminishing role of crude oil with a high content of sulfur in the oil industry;
- environmental pressures in the marine fuel industry and the power generation sector;
- increase in the use of unconventional oil sources for electricity generation and heating such as shale oil, heavy oil, compact oil, and tar sand;
- increased production of both mature and frontier fields (World Energy Resources, 2016, used with the permission of the World Energy Council, www.worldenergy.org, 2016).

The replacement of crude oil and coal by other cleaner energy sources for electricity generation and heating with fewer adverse effects on the environment and population is not yet imminent and is not expected to reach more than 5% over the next five years. During this time, all types of oil (conventional and unconventional oil) will continue to be used for electricity generation and heating in several countries, due to significant crude oil reserves reported by several states in different regions. The recovery of unconventional oil represents 30% of the global reserves of recoverable crude oil, and shale oil resources contain at least three times more oil than conventional oil reserves, which are estimated to be around 1,2 trillion barrels (World Energy Resources, 2016, used with the permission of the World Energy Council, www.worldenergy.org, 2016). The replacement of crude oil or coal by any other energy source for electricity generation and heating is, for several countries, not yet an economical alternative. It is likely that this situation will not change any time soon.

The implementation of the energy policy adopted by several countries to completely decarbonize their national energy system is expected to have significant consequences for the future income of fossil–fuel extraction companies. Countries that export this type of fuel will also be affected, but the risks vary according to the kind of fossil fuel in question. For example, the main threat in the coal sector is concentrated in the operation of coal-fired power plants (for which carbon capture and storage became an essential asset protection strategy). But the key risk in the mining sector, which is much less capital intensive, is in the employment of the labor force. For this reason, it is crucial that fossil fuel exporting countries take steps to reduce their vulnerabilities by limiting their dependence on fossil fuel income, as Saudi Arabia does with its comprehensive "Vision 2030" reform program.

In the case of crude oil, there is no reason to suppose a general stranding of oil assets in an upward direction, provided that governments give clear signals of their intention and apply coherent energy policies to that end. Investment in the development of new upstream projects is an essential component of a lower-cost transition since the decrease in the production from existing crude oil fields is much higher than the expected fall in the crude oil demand. However, the risks would increase drastically in case of sudden changes in energy policies, or other circumstances that lead companies to invest in energy demand that does not materialize.

It is important to single out the fact that the "North American crude oil market has undergone a significant transformation in recent years with the emergence of light crude oil production from shale plays located throughout the US" (Crude Oil: Forecast, Markets, and Transportation, 2017).

The US being the primary destination for Canada's crude oil supplies, significant opportunities are still available that would increase the volumes that could be sold to this market. "However, it is important to know that the growth in Canadian production is forecast to be mainly heavy crude oil from oil sands, while US domestic production will be predominately light crude oil" (Crude Oil: Forecast, Markets, and Transportation, 2017). Despite the investment in technology already made in US refineries to process heavy crude oil, they prefer to rely on cheaper feedstock. However, US domestic light crude oil supplies are no longer captive to US refineries since the removal of restrictions on US crude oil exports approved by the government in December 2015 (Crude Oil: Forecast, Markets, and Transportation, 2017).

In summary, the following can be stated related to the current world oil sector and its perspective:

1. Crude oil remains the leading fuel in the world, accounting for 32,9% of the total world energy consumption.
2. Emerging economies now account for 58,1% of global energy consumption and global demand for liquids will continue to grow (BP Statistical Review of World Energy 2016, 65th edition, 2016). In 2015, a rare increase in the use of energy in the EU of 1,6% offsets the decline in energy consumption in the US of 0,9% and in Japan of 1,2%, where energy utilization fell to the lowest level since 1991 (World Energy Resources, 2016 used with the permission of the World Energy Council, www.worldenergy.org, 2016).

3. The growth of world population and energy consumers in Asia will cause an increase in world oil demand in the coming decades, but particularly in that region. The core rise in oil consumption will come from the transport sector.

4. Fluctuations in crude oil prices lately have been neither unexpected nor unprecedented.

5. The primary drivers of crude oil price changes have been the gradual increase in the additional storage capacity of OPEC, and the appearance in the market of new non-OPEC oil suppliers, especially of US light oil.

6. New and increased use of different technologies (high-pressure, high-temperature drilling, multistage fracking, development in flow assurance for mature fields, greater sophistication in well-simulation techniques, reservoir's modeling, and 3-D seismic technologies) are having a positive impact on safety possibilities during the manipulation of crude oil by a different sector of the oil industry (World Energy Resources, 2016, used with the permission of the World Energy Council, www.worldenergy.org, 2016).

The Crude Oil Sector in the United States

The US was self-sufficient in the supply of crude oil for many years after its discovery in 1859. In 1994, the US began to import more crude oil than it produced. In 2015, the US imported 24% of the total crude oil consumed in the country. In that year, the six main crude oil and other petroleum products supplying countries were Canada, Saudi Arabia, Venezuela, Mexico, Colombia, and Russia. However, it is important to highlight that according to several experts' opinions, the US will become a larger producer of crude oil than Saudi Arabia in the coming years as a result of the exploitation of tight oil deposits. If this projection is met, then an enormous transformation of the US within the global oil markets will occur and will have a significant impact on the overall global oil balances, given the reduced need of oil in the US. According to Nyquist (2018), director of McKinsey's Houston office, "the US could be net exporters of energy by 2020. It's a significant shift in the way we think about energy security, and the way we think about the impact of energy prices on our economy."

The US has enormous reserves of conventional and non–conventional fossil fuels. The responsible use of these resources has allowed many of its factories and furnaces to operate uninterruptedly, heat millions of homes across the country, build motorways and hospitals, only to mention some of the things that can be done using all fossil fuels available in the country. Rationally using these resources will allow the conservation of large quantities of these unexploited resources for their use by future generations. Access to these energy resources and the discovery of new technologies to use them safely and cleanly for the benefit of humanity remain a constant challenge. Government decisions about the use of these energy resources have placed them outside of the exploitation by the US industry, one of the main reasons for the increase in US dependence on other countries for the supply of these energy resources in the last years (Oil-IER, 2017).

The US has important reserves in crude oil, natural gas, and coal, and uses different renewable energy sources as well as nuclear energy as primary energy sources for electricity generation and heating. In 2016, the total US primary energy consumption was about 97,4 quadrillion (10^{15} or 1.000 trillion) Btu. In 2016, and according to EIA sources, the shares of total primary energy consumption for the five energy-consuming sectors were:

- electric power: 39%;
- transportation: 29%;
- industrial: 22%;
- residential: 6%; and
- commercial: 4%.

The electric power sector produced most of the electricity in the US and consumed 39% of the total power generated in the country. The other four areas consume 61% of the whole power generated in the country, the transport sector being the one that consumes the highest percentage within that group. The pattern of fuel use varies widely by sector. For example, crude oil provides about 92% of the energy used for transportation, but only 1% of the energy used for electricity generation and heating. It is probable that the percentage of the participation of crude oil in the electricity generation and heating in the US will continue to be very small during the coming years.

Crude Oil Reserves in the United States

According to the US Crude Oil and Natural Gas Proved Reserves Year-End 2016 report (EIA 2018), between the end of 2014 and the end of 2015, the US proven crude oil and lease condensate reserves decreased by 11,8%, falling from 39.900 billion barrels in 2014 to 35.200 billion barrels in 2015; this means a decrease of 4,7 billion barrels. In 2016, total US crude oil and lease condensate proved reserves remained at 35.200 billion barrels and US crude oil proved reserves reached 32.773 million barrels (see Figs. 2.1 and 2.2).

It is important to highlight that proved reserves of crude oil and lease condensate increased in four of the top seven US crude oil reserves states in 2016. These states are Texas (1,8 million barrels), North Dakota (0,4 million barrels), Oklahoma (0,3 million barrels), and New Mexico (0,2 million barrels).[3]

In the specific case of tight oil, the US proved reserves of this type of oil, according to EIA sources, were estimated to be 58 billion barrels or 16,8% of the world total tight oil proved reserves.

By states, the changes in crude oil and lease condensate proved reserves in 2016 with respect to 2015, according to EIA sources, are the following:

- Texas had the most substantial net increase in crude oil and lease condensate proved reserves (941 million barrels), an increase of 7% concerning the previous year[4].
- Oklahoma had the second-largest net increase in crude oil and lease condensate proved reserves (386 million barrels), an increase of 23% concerning the previous year.
- New Mexico had the third-largest increase in crude oil and lease condensate proved reserves (74 million barrels), an increase of 5% with respect to the previous year.
- Alaska had the most significant net decline of crude oil and lease condensate proved reserves of all US states, a 25% drop of 530 million barrels.
- California and the Federal Offshore Pacific had the next largest net declines in crude oil proved reserves (402 million barrels and 203 million barrels, respectively) (US Crude Oil and Natural Gas Proved Reserves Year-End 2016; 2018).

The evolution of crude oil proved reserves in the US during the period 2010—16 is shown in Fig. 2.2.

---

[3] The top seven US crude oil and lease condensate proved reserves states, in 2016, were the following: Texas, North Dakota, California, Alaska, Oklahoma, New Mexico, and the Gulf of Mexico.

[4] "In 2015, Texas proved crude oil reserves had declined more than in any other state (1.001 million barrels) after operators revised their proved reserves downward in response to the dramatic drop in crude oil prices at the end of 2014 from an average of about US\$95 per barrel to US\$50 per barrel" (US Crude Oil and Natural Gas Proved Reserves Year-End 2016, 2018).

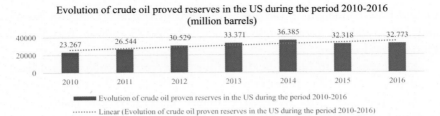

Figure 2.2 Evolution of crude oil proved reserves in the US during the period 2010—16 (million barrels). *(Source: US Crude Oil and Natural Gas Proved Reserves Year-End 2016 (2018).)*

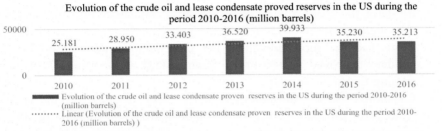

Figure 2.3 Evolution of crude oil and lease condensate proved reserves in the US during the period 2010—16 (million barrels). *(Source: US Crude Oil and Natural Gas Proved Reserves Year-End 2016 (2018).)*

According to Fig. 2.2, the crude oil proved reserves in the US increased by 40,9% during the period 2010—16, rising from 23.267 million barrels in 2010 to 32.773 million barrels in 2016. The peak in the level of the US crude oil proved reserves was reached in 2014. It is expected that the US crude oil proved reserves would continue to grow during the coming years, particularly in the case of unconventional oil, as a result of new discoveries.

If we considered the US crude oil and lease condensate proved reserves, then the evolution of these reserves during the period 2010—16 can be seen in Fig. 2.3.[5]

According to Fig. 2.3, the crude oil and lease condensate proved reserves in the US increased by 39,8% during the period 2010—16, rising

---

[5] According to BP Statistical Review of World Energy 2018 report, the US crude oil reserves at the end of 2017 reached 50.000 million tons.

from 25.181 million barrels in 2010 to 35.213 million barrels in 2016.[6] The peak in the level of the crude oil and lease condensate proved reserves in the US within the period considered was reached in 2014. It is predictable that the US crude oil and lease condensate proved reserves would continue to grow slowly during the coming years, particularly in the case of unconventional oil, despite the reduction of these reserves during the period 2014—16.

In general, when crude oil prices fall or operating costs increase, the life of a producing well can be shortened by reducing the remaining crude oil proved reserves that are economically recoverable and can cancel or postpone planned oil wells construction. This action may reduce the estimate of crude oil proved reserves, even if production is increasing (generally due to a large number of new wells drilled and completed before the increase in price/cost decline). Therefore, changes in annual crude oil production may be delayed concerning the year in which changes in crude oil proved reserves are reported. In principle, a drilled and completed crude oil well is required to have proved reserves. However, the crude oil proved reserves can be assigned to a planned well, based on the surrounding wells and the geological and engineering data of these wells (if the scheduled well will be drilled within five years). It is important to highlight that crude oil production cannot increase or maintain a certain level for a long time without continuous development or improved methods of crude oil recovery. There should be no doubt that crude oil production will inevitably decrease in the absence of constant improvement in the wells used.

Estimates of crude oil reserves change from year to year as a result of four reasons, which are:
- New crude oil deposits are found.
- Existing oil fields are evaluated more thoroughly.
- Existing crude oil reserves are beginning to be exploited.
- There are movement in prices and improvement in the technologies used in the production of crude oil.

According to EIA sources, it is very likely that crude oil reserves will be revised downwards in the next EIA report, but probably not in the same manner as in the previous release. Low crude oil prices have reduced the

---

[6] "US crude oil and lease condensate proved reserves decreased by seven million barrels in 2016 as net additions (mostly extensions and discoveries) were virtually the same as annual production" (US Crude Oil and Natural Gas Proved Reserves Year-End 2016, 2018).

Table 2.1 The US Proved Crude Oil Reserves and Reserves Change (31.12.2016).

	Crude Oil (million barrels)
US crude oil proved reserves	32.773
Total extensions and discoveries	2.794
Net revisions (revision increases minus revision decreases)	17
Sales	1.125
Acquisitions	1.460
Estimated production	2.953
US crude oil proved reserves at December 31, 2015	32.318
Percentage change in the US proved reserves	1,7%

Source: EIA database (2016).

drilling of new wells in recent years. Although low crude oil prices do not necessarily diminish estimates of technically recoverable crude oil reserves, the calculation of crude oil proved reserve is sensitive to changes in crude oil prices (Table 2.1).

The trend in the evolution of crude oil and lease condensates proved reserves in the US is very clear. It shows a decline during the period 1970—2008. During the period 2009—15, an increase in the US crude oil and lease condensates proved reserves was registered. Since 2015, once again a decline in crude oil and lease condensates proved reserves throughout the country was recorded.

Finally, it is important to highlight that, contrary to the idea that the US is running out of oil, the country continues to be rich in this type of fossil fuel. It is true that some crude oil sources have been too expensive for their use in the past, but now they are cheap, as a result of the use of new drilling techniques. Shale oil, for example, has experienced the same renaissance as shale gas due to hydraulic fracturing and directional drilling.

In summary, the following can be stated: recent discoveries of new reserves of crude oil and lease condensates in the US totaled 3.247 million barrels. Geographically, the most significant reserves were found in Texas, North Dakota, and Oklahoma. Texas reported discoveries for 1.400 million barrels, North Dakota, for 0,600 million barrels, and Oklahoma for 0,4 million barrels. Total crude oil discoveries in the federal Gulf of Mexico were 108 million barrels, 20 million barrels of which came from new field discoveries. In 2015, all US new field discoveries were in the Gulf of Mexico. In 2016, the findings and extensions reached 3.204 million barrels, which is 1,4% lower than the one reported in 2015.

Crude Oil Production in the United States

The US oil industry has been an essential industry since its discovery in Titusville, Pennsylvania in 1859. The petroleum industry includes activities such as exploration, production, processing (refining), transport, and marketing of petroleum products.

The US is the fourth largest crude oil producer in the world (after Venezuela, Saudi Arabia, and Russia), producing 8.857 million barrels of crude oil per day in 2016 (EIA database, 2017). The main crude oil producing areas in the US during the period 2011−16 are included in Table 2.2.

Table 2.2 Evolution of Crude Oil Production in the US by Regions During the Period 2011−16 (thousand barrels per day).

	2011	2012	2013	2014	2015	2016
United States	5.643	6.497	7.466	8.753	9.408	8.857
PADD 1	22	26	40	47	49	44
Florida	6	6	6	6	6	5
New York	1	1	1	1	1	1
Pennsylvania	9	12	15	19	19	17
Virginia	0	0	0	0	0	0
West Virginia	6	7	18	21	23	20
PADD 2	818	1.132	1.408	1.734	1.893	1.679
Illinois	25	27	26	26	26	24
Indiana	5	6	7	7	6	5
Kansas	114	120	128	136	125	104
Kentucky	6	9	8	9	8	7
Michigan	19	20	21	20	18	15
Missouri	0	0	1	1	0	0
Nebraska	7	8	8	8	8	6
North Dakota	418	662	856	1.081	1.177	1.033
Ohio	13	14	22	41	73	60
Oklahoma	206	259	326	399	447	420
South Dakota	4	5	5	5	5	4
Tennessee	1	1	1	1	1	1
PADD 3	3.252	3.776	4.375	5.200	5.647	5.457
Alabama	23	26	29	27	27	22
Arkansas	11	13	13	13	12	15
Louisiana	189	193	197	188	172	154
Mississippi	66	67	67	67	68	56
New Mexico	196	234	281	342	404	399

Continued

Table 2.2 Evolution of Crude Oil Production in the US by Regions During the Period 2011—16 (thousand barrels per day).—cont'd

	2011	2012	2013	2014	2015	2016
Texas	1.450	1.978	2.534	3.166	3.448	3.213
Federal offshore PADD 3	1.317	1.266	1.255	1.397	1.515	1.598
PADD 4	396	449	531	664	753	662
Colorado	108	136	181	262	336	317
Idaho	0	0	0	0	0	1
Montana	66	72	80	82	78	63
Utah	72	83	96	112	102	83
Wyoming	150	158	174	209	237	198
PADD 5	1.154	1.114	1.113	1.109	1.067	1.015
Alaska	561	526	515	496	483	490
South Alaska	10	11	15	18	18	15
North slope	551	515	500	478	465	474
Arizona	0	0	0	0	0	0
California	537	539	546	561	551	508
Nevada	1	1	1	1	1	1
Federal offshore PADD 5	54	48	51	51	31	17

Notes: Year-to-date totals include revised monthly production estimates by state published in Petroleum Navigator. Crude oil production quantities are estimated by state and summed to the PADD and the US level. State production estimates reported by EIA usually are different from data published by state agencies. Totals may not equal sum of components due to independent rounding.
Source: EIA database (2017).

It is important to highlight that, in 2015, US crude oil production increased to 9.408 million barrels per day, the highest annual crude oil production since 1972. This level of production represented an increase of 90% over 2008 production of five million barrels per day (Petroleum in the United States, 2018). Monthly, the output of crude oil in the US reached a maximum of 9.630 million barrels per day in April 2015 but then decreased to 8.840 million barrels per day in 2016 (see Table 2.2) due to a drop in crude oil prices. In 2017, crude oil production increased by 6% concerning the level reached in 2016 (9.367 million barrels per day).

Between 2010 and 2015, there was a boom in shale oil production in the US. Considering the high level of imports of crude oil from the US, which reached the amount of more than 10 million barrels per day in recent years, the production of shale oil provided greater energy security by helping to reduce crude oil imports, and to increase domestic production. However, and despite this boom in the production of shale oil in the US, in 2016, the production of crude oil from the US decreased by 6,1%

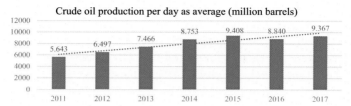

Figure 2.4 Crude oil production per day as average in the US during the period 2011–17. *(Source: EIA database (2017).)*

(see Fig. 2.4) concerning 2015. However, in 2017, the output of crude oil in the US increased once again (5,9% with respect to 2016).

The total crude oil production per year in the US during the period 2011–17 is shown in Fig. 2.5.

According to Fig. 2.5, the production of crude oil in the US increased by 66,7% during the period 2011–15, rising from 2.059.640 thousand barrels per year in 2011 to 3.434.028 thousand barrels per year in 2015, but decreased by 5,8% from 2015 to 2016, the first reduction in the last seven years, and once again increased by 5,7% during the period 2016–17. Considering the whole period 2011–17, the US production of crude oil rose by 66%. It is expected that the production of crude oil in the US will continue to grow during the coming years, particularly of tight oil if new discoveries are found.

On the other hand, it is important to highlight that the highest US production of crude oil was 9,6 million barrels per day in June 2015, the peak level reached in the crude oil production in the US history, but one year later fell to 8,4 million barrels per day in July 2016 (the lowest level since May 2014). The EIA has estimated that US crude oil production could average 9,3 million barrels per day in 2017 and 9,8 million barrels per day as average in 2018. If these estimates are correct, then the US crude oil

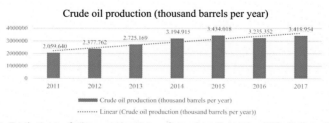

Figure 2.5 Evolution of the production of crude oil in the US during the period 2011–17. *(Source: EIA database (2018).)*

production could reach a new annual record in 2018. Crude oil production in the US averaged 9,4 million barrels per day in 2015 and 8,9 million barrels per day in 2016; this represented a 5,4% decrease between 2015 and 2016 (Kristopher, 2017). It is forecasted that US crude oil production will reach 10,1 million barrels per day in December 2018,[7] that is, 0,9 million barrels per day more than the level registered in June 2017, and 1,4 million of barrels per day more since the end of 2016[8] (International Energy Outlook 2016 with Projection to 2040, 2016).

Undoubtedly, crude oil, natural gas, and coal are the types of energy sources that have dominated the US energy mix for more than a century. However, there have been several recent changes in the US energy production, which has affected the structure of the country's energy mix. In the case of crude oil, the main difference is as follows: crude oil production generally decreased each year between 1970 and 2008. In 2009, the trend reversed, and production began to rise. More cost-effective drilling and production technologies helped to boost US crude oil production, especially in Texas and North Dakota. However, in 2016, US crude oil production was 9% lower than the output registered in 2015, mainly because of lower global crude oil prices[9] (Swier, 2018).

---

[7] In its latest Short-Term Energy Outlook (May 2018) report, the US EIA forecasts US crude oil production to average 10,7 million barrels per day in 2018 from 9,4 million barrels per day in 2017 (an increase of 13,8%) and expected to average 11,9 million barrels per day in 2019 (an increase of 11,2%). In the current outlook, EIA forecasts US crude oil production will end 2019 at more than 12 million barrels per day.

[8] In 2016, according to EIA sources, energy produced in the US was equal to about 86% of US energy consumption. The difference between production and consumption was mainly in net imports of petroleum. The three major fossil fuels—petroleum, natural gas, and coal—accounted for most of the nation's energy production in 2016: Natural gas: 33%; petroleum (crude oil and natural gas plant liquids or lease condensate): 28%; coal: 17%; renewable energy: 12%; and nuclear energy: 10%.

[9] "Oil and natural gas prices have fallen dramatically thanks to the fracking revolution and increased production. This increased production, however, has occurred in spite of — rather than because of — Obama administration policies. Fortunately for American consumers, increased production on privately owned and state-owned lands has more than compensated for the Obama administration increasing the percentage of federal-owned lands rendered off-limits to oil and gas production". It is expected that the Trump administration open up more federal lands to energy production for private companies, which will further increase domestic oil and natural gas production. This increase in the domestic production of oil "will in turn lower energy prices, increase royalty payments to offset our national debt, and bolster the American economy" (Taylor, 2016).

The increase in US crude oil production has occurred on both private and public lands. However, oil production on federal lands has decreased recently mainly due to the measures adopted by the previous Obama administration that have increased the time for processing of licenses for the extraction of crude oil on certain public lands and limited access to public lands rich in oil. Crude oil can be accessed on these lands with new technologies and only after government approval (Oil-IER, 2017).

Unconventional Oil in the United States

Unconventional oil is petroleum produced or extracted using techniques other than the conventional (oil well) method. Oil industries and governments across the globe are investing in unconventional oil sources due to the increasing scarcity of conventional crude oil reserves at the world level (Gupta). "Unconventional oil includes:

- oil shales;
- oil sands–based synthetic crudes and derivative products;
- coal–based liquid supplies; and
- liquids that are arising from chemical processing of gas" (Gupta). Other sources include:
- extra–heavy crude oil;
- natural bitumen (oil sands);
- kerogen oil;
- liquids and gases arising from chemical processing of natural gas (GTL);
- coal–to–liquids (CTL), and additives (World Energy Outlook 2001, 2001).

In general, conventional oil is easier and cheaper to extract than unconventional crude oil. However, it is important to highlight that these two categories do not remain fixed, and one type of unconventional oil can be reclassified as conventional oil over time, as a result of changes in the economic and technological conditions prevailing in a given time.

Tight Oil in the United States

It is very likely that the US energy future is strongly influenced by the use of new technologies in the field of unconventional oil, particularly in the case of tight oil. Procedures such as horizontal drilling and hydraulic fracturing have allowed the extraction of low permeability crude oil deposits that were classified as unproductive or marginal at the end of the 1990s. The advances in the use of these new techniques have caused a tight oil boom that has significantly altered the global oil markets. In fact, in the midst of a reasonably flat demand for crude oil, companies are now looking forward to exporting excess supply of crude oil from the US, an action unthinkable a decade ago (Clemente, 2015). The availability of abundant and cheap tight oil has been an essential injection into the US economy.

According to the Annual Energy Outlook 2018 report, the US tight oil production is expected to increase through the early 2040s, when it is supposed to surpass 8,2 million barrels per day and to account for nearly 70% of total US oil production projection for that period. It is also predictable that the US tight oil production will remain relatively constant through 2050 as development moves into less productive areas. Development of tight oil resources is more sensitive than non–tight oil to different assumptions of future crude oil prices, drilling technology, and resource availability, but tight oil will remain the largest source of US crude oil production during the coming years (Today in Energy, 2018).

Tight Oil Production in the United States

According to the International Energy Outlook 2016 with Projection to 2040 report, "US tight oil production, which reached 4,6 million barrels per day in May 2015, decreased to 4,3 million barrels per day in February 2016". Tight oil production has proven to be more unaffected by low prices of crude oil that many market experts had projected.

The EIA estimates that, in 2017, about 4,67 million barrels per day of crude oil were produced directly from tight oil resources in the US, an increase of 8,6% with respect to February 2016. This level of production was equal to about 50% of total US crude oil production in 2017. According to EIA projections, in 2020, about 18% of the US total oil production will come from tight oil and, by 2035, this type of oil will account for 20,5% of the total oil production in the country. Tight oil output in the US is expected to increase until 2029 and then begin to decline until 2035. However, these projections seem conservative relative to other market analysts who forecast US tight oil production between 3 and 4 million barrels per day (Shale Oil: The Next Energy Revolution, 2013). The main tight oil and shale gas regions in the US are included in Fig. 2.6.

It is important to single out that "economic production from tight oil formations requires the same hydraulic fracturing and often uses the same horizontal well technology used in the production of shale gas. While sometimes called "shale oil," tight oil should not be confused with oil shale,

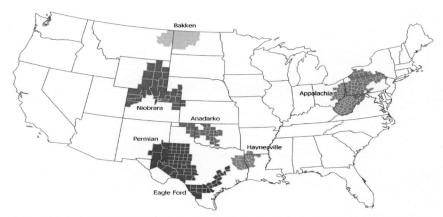

Figure 2.6 Key tight oil and shale gas regions in the US. *(Source: EIA (https://www.eia. gov/petroleum/drilling).)*

which is shale rich in kerogen, or shale oil, which is oil produced from oil shales" (Reinsalu and Aarna, 2015)[10].

On the other hand, "tight oil production, which is characterized as light sweet crude oil and lease condensates, is expected to continue to fuel the bulk of future US oil production growth." As shown in Fig. 2.6, "this production growth is replacing light oil imports, mostly African crude oil, at an increasing clip. Between January 2010 and January 2014, the light oil imports dropped by nearly two-thirds to 790.000 barrels per day for the US. In the Gulf Coast (PADD 3), light oil imports are currently only about 245.000 barrels per day, and a portion of these are for lubricant oil manufacturing (requiring a specific crude oil quality)" (Annual Energy Outlook 2014, 2014).

According to Fig. 2.7, the total crude oil production in the US increased during the period 2011−14, due to the growth in the conventional crude oil production. At the same time, the imports of light crude oil decreased during the same period.

Finally, it is important to emphasize that tight oil producers are rapidly developing new hydraulic fracturing or fracking techniques that could

---

[10] "Based on US shale production experience, the recovery factors for shale gas generally ranged from 20% to 30%, with values as low as 15% and as high as 35% being applied in exceptional cases. Because of oil's viscosity and capillary forces, oil does not flow through rock fractures as easily as natural gas. Consequently, the recovery factors for shale oil are typically lower than they are for shale gas, ranging from 3% to 7% of the oil in-place with exceptional cases being as high as 10% or as low as 1%" (Annual Energy Outlook, 2015, 2015).

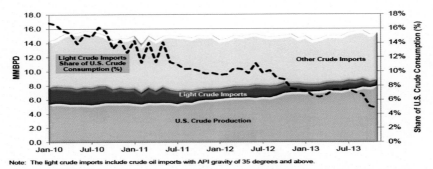

Note: The light crude imports include crude oil imports with API gravity of 35 degrees and above.

Figure 2.7 Historical US crude oil production and light crude oil imports. *(Source: Crude Oil Production (2014); EIA; http://www.eia.gov/dnav/pet/pet_crd_crpdn_adc_mbbl_m.htm; 2014.)*

increase the US crude oil reserves in the future, particularly tight oil reserves. But at the same time, producers face significant challenges in delivering crude oil to the market at an adequate price to generate a return commensurate with the investments and expenses made. Midstream players have been moving to provide services to both producers and refiners. These services include rail facilities for loading and unloading crude oil, new terminals, and pipelines to bring crude oil to the market, as well as facilities to export more and more oil products to foreign markets.

These changes are not only transforming the US oil and petrochemical industries but, at the same time, stimulating the economy by boosting investment, employment, and GDP growth at all points along the oil supply chain. Along with increased Canadian crude oil production, the growth of domestic crude oil and condensate production has dramatically reduced the dependence on oil from outside the North American region.

The Shale Oil Revolution and Its Impact on Crude Oil Production in the United States

Throughout the world, vast quantities of crude oil and natural gas are trapped in non-permeable shale rock. In recent years, technological advances in hydraulic fracturing or fracking and horizontal drilling have made recovery of much of this type of oil and gas economically feasible. The most significant US shale oil deposits[11] are located in the Bakken in North

---

[11] According to public information available on tight oil, no other nation in the world is as rich in oil shale as the US, with over two trillion barrels of tight oil (Oil-IER, 2017).

Dakota; Barnett, Eagle Ford, and Permian in Texas; Marcellus in the east, especially Pennsylvania and Ohio; and Niobrara in Wyoming and Colorado (see Fig. 2.6). Some areas contain more natural gas than crude oil; others contain more crude oil than natural gas. Thanks to shale oil, economically recoverable US natural gas and crude oil reserves are much higher than they were thought to be just a few years ago[12] (Moving Crude Oil by Rail, 2014). The largest deposits of American oil shale are situated in Colorado, Utah, New Mexico, and Wyoming. Though the extraction poses some technical challenges, US oil shale in-place resources contain the energy equivalent of over two trillion barrels of crude oil.

According to the US Rail Crude Oil Traffic (2017) report, much of the recent increase in the US crude oil production has been in North Dakota, where the crude oil production increased from an average of 81.000 barrels per day in 2003 to 1,2 million barrels per day in 2015; this means an increase of 18,4 times. However, from 2015 to 2016 the crude oil production in North Dakota decreased to 1 million barrels per day; this means a decrease of 16,7%. Despite this decrease, North Dakota is the second-largest oil producing state in the entire country. The production of crude oil in Texas has skyrocketed since 2009, reaching an average of 3,5 million barrels per day in 2015. In 2016, the crude oil production in Texas registered a decrease of 8,6% or a total of 3,2 million barrels per day.

It is difficult to identify all economic benefits associated with the growth in the national production of crude oil in the US as a result of the discovery of new shale oil deposits. Some of the main benefits are, among others:

- lower dependence on imports of crude oil from external sources in a world that is not politically stable and whose interests do not necessarily correspond to those of the US;

---

[12] When considering the market implications of abundant oil shale resources, it is important to distinguish between a technically recoverable resource, and an economically recoverable resource. In the first case, is spoken of oil and natural gas that could be produced with current technology, regardless of oil and natural gas prices and production costs. In the second case, is spoken of resources that can be profitably produced under current market conditions. "The economic recoverability of oil and gas resources depends on three factors: the costs of drilling and completing wells, the amount of oil or natural gas produced from an average well over its lifetime, and the prices received for oil and gas production. Recent experience with shale gas and tight oil in the United States and other countries suggests that economic recoverability can be significantly influenced by above-the-ground factors as well as by geology" (Technically Recoverable Shale Oil and Shale Gas Resources: Canada, 2015).

- lower vulnerability to oil crises that in the past have caused immense damage to the US economy;
- new and better employment opportunities and economic development for communities throughout the country;
- billions of dollars in new tax revenues; and
- reduction of billions of dollars every year in the US trade deficit.

The increase in the US crude oil production, combined with a relatively weak global demand for it, due to the difficult economic situation that many countries are facing, led to an oversupply of crude oil and a sharp drop in oil prices since June 2014. In January 2016, the spot prices of crude oil were more than 70% lower than in June 2014. Some US producer who could not extract crude oil profitably to that price level was forced to leave the market, which helps explain why crude oil production in 2016 was 8,6% lower than in 2015. Continuous technological advances have allowed many US crude oil producers to reduce their extraction costs, achieving a renewed growth in the US crude oil production (US Rail Crude Oil Traffic, 2017).

Crude Oil and Other Liquid Fuels Consumption in the United States

According to the Short-Term Energy Outlook (2017) document, total consumption of crude oil and other liquid fuels in the US is forecasted to average 19,9 million barrels per day in 2017, which would represent an increase of 310.000 barrels per day or 1,6% more compared to the level reached in 2016. Also, it is forecasted that crude oil consumption will grow by 360.000 barrels per day in 2019 or 1,8% more than the level reached in 2018. It is anticipated that growth in both years will be led by higher consumption of oil, hydrocarbon gas liquids, and distilled fuel.

It is important to highlight that almost all the crude oil that is produced within the country or imported from a third country is refined to produce petroleum products such as gasoline, diesel fuel, heating oil, and jet fuel, which are then consumed inside of the US. Liquids produced from natural gas processing are also absorbed inside the US as petroleum products. In 2016, the US used up a total of 7,19 billion barrels of petroleum products, an average of approximately 19,63 million barrels per day. "In 2017, the US consumed a total of 7,26 billion barrels of petroleum products, an average of about 19,88 million barrels per day" (How Much Oil is Consumed in the United States, 2018), and an increase of 1,3% concerning 2016. The share of crude oil consumption by sectors within the total US energy consumption is shown in Fig. 2.8.

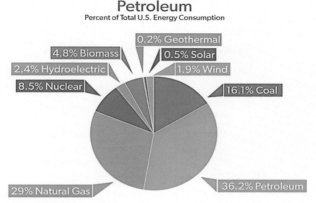

Figure 2.8 Percentage of total US energy consumption by sectors. *(Source: EIA database (2016).)*

According to Fig. 2.8, the consumption of petroleum within the US total energy consumption was, in 2016, of 36,2%, followed by natural gas with 29%, and coal with 16,1%, all of them classified as conventional energy sources. It is expected that this situation will not change significantly during the coming years.

Corresponding to Fig. 2.9, in 2016, fossil fuels generated 81% of the total energy consumed in the country,[13] followed by renewable energy with 10%, and nuclear power with 9%. By type of energy source, biomass, including biomass waste (5%), biofuels (22%), and wood (19%), are the primary sources of energy with 46% of the total, followed by petroleum with 37%, natural gas with 29%, hydroelectric with 24%, wind with 21%, and coal 15%.

As it was said before, the use of crude oil for electricity generation and heating is very low today, and it is expected that this level will not change during the coming years. The evolution in the use of oil for electricity generation and heating in the US during the period 2010—17 is included in Fig. 2.10.

On the basis of the data included in Fig. 2.10, the following can be stated: the use of oil for electricity generation and heating in the US decreased by 43% during the period considered, falling from 37.062

---

[13] "The crude oil demand was strong in the petrochemical industry in the US, which has benefited from the fact that rising supply has left American crude oil prices lower than those in many other countries" (Norris, 2014).

U.S.energy consumption by energy source, 2016

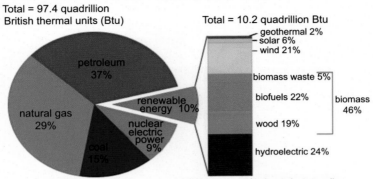

Note: Sum of components may not equal 100% because of independent rounding.

Figure 2.9 US energy mix in April 2017. *(Source: EIA database (2017).)*

Evolution of the electricity generation in the US using crude oil as fuel (thousands MWh)

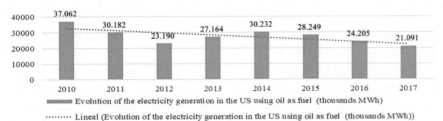

Figure 2.10 Evolution of the electricity generation using crude oil as fuel in the US during the period 2010−17. *(Source: EIA database (2018).)*

thousand MWh in 2010 to 21.091 thousand MWh in 2017. In that year, the electricity generation and heating in the US using crude oil as fuel representeds only 0,5% of the total electricity generated in the country. It is likely that this percentage will continue to be very small during the coming years. Within the period considered, the peak in the electricity generation and heating using crude oil as fuel was reached in 2010. Since that year, except the years 2013 and 2014, electricity generation and heating using crude oil as fuel decreased systematically. It is expected that this trend will continue without change during the coming years, and the participation of crude oil in the electricity generation and heating in the US will be lower than in 2017.

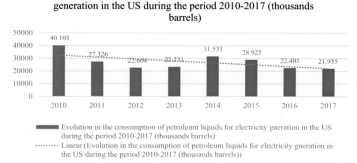

Evolution in the consumption of petroleum liquids for electricity generation in the US during the period 2010-2017 (thousands barrels)

Figure 2.11 Evolution of the consumption of petroleum liquids for electricity generation and heating in the US during the period 2010—17. *(Source: EIA 2018 database for the EIA energy conference of 2018.)*

The evolution of the consumption of petroleum liquids for electricity generation and heating in the US during the period 2010—17 is included in Fig. 2.11.

According to Fig. 2.11, the use of petroleum liquids for electricity generation and heating in the US during the period 2010—17 decreased by 45,4%, falling from 40.103 thousand barrels in 2010 to 21.935 thousand barrels in 2017. According to the new energy policy adopted by the Trump administration, the use of petroleum liquids for electricity generation and heating will continue to decrease during the coming years, mainly in favor of the increased use of natural gas and renewables. The peak in the use of petroleum liquids for electricity generation and heating in the US within the period considered was reached in 2010. During the period 2010—13, the use of petroleum liquids for electricity generation and heating decreased by 48%; during the period 2014—17 decreased by 30,5%. It is forecast that the participation of liquids oil in the electricity generation sector of the US will be minimal during the coming years.

Finally, it is essential to highlight the following: currently, the US is the largest OECD consumer of fuels and will remain so until 2040. However, it is expected that the use of liquid fuels in the US transport sector will decrease during the period 2012—40 as a result of lower fuel consumption in light vehicles. The decrease is expected to be moderate due to higher fuel consumption in the case of heavy cars, airplanes, and marine vessels (International Energy Outlook 2016 with Projection to 2040, 2016).

Crude Oil Imports by the United States

The US is importing crude oil from different countries and regions. The country from where the US imports most of its crude oil is Canada with 40% of the total, followed by some OPEC members, such as Saudi Arabia, Venezuela, Iraq, Nigeria, Ecuador, and Kuwait with 31%, Mexico 8%, Russia with 4%, and several other countries with 17% (See Fig. 2.12).

The evolution of the imports of crude oil by the US during the period 2011—17 is shown in Table 2.3.

According to Table 2.3, in 2011 the US imported crude oil from 43 countries. The leading suppliers were Canada, Saudi Arabia, and Mexico, in addition to the countries included within the Persian Gulf. In 2016, the US imported crude oil from 37 countries; 14% less than in 2011. In 2017, the US imported crude oil from 38 countries, one more than in 2016. The leading suppliers of crude oil to the US in 2017 were Canada, Saudi Arabia, Venezuela, and Mexico, in addition to the countries included within the Persian Gulf.

It is important to highlight that since 2013, the production of crude oil in the US (7,466 million barrels per day) was slightly lower than crude oil imports (7,73 million barrels per day) and this trend continued until 2017. During the period 2013—17, the production of crude oil in the US continued to grow and reached the amount of 9,367 million barrels per day (an increase of 25,5%). The imports of crude oil also increased to 7,912 million barrels per day (an increase of 2,4%). It is predictable that the production of crude oil in the US will continue to increase for at least the next few years, reducing the need to import crude oil from other countries.

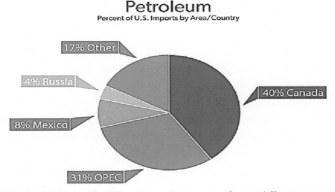

Figure 2.12 Percentage of US crude oil imports from different countries and areas. *(Source: EIA sources.)*

Table 2.3 Evolution of the Imports of Crude Oil by the US During the Period 2011–17 (thousand barrels per day).

	2011	2012	2013	2014	2015	2016	2017
All Countries	8.935	8.527	7.730	7.344	7.363	7.850	7.912
Persian Gulf	1.849	2.140	1.994	1.851	1.487	1.738	1.710
OPEC[a]	4.209	4.031	3.493	3.005	2.673	3.180	3.112
Algeria	178	120	29	6	3	51	66
Angola	335	222	201	139	124	159	129
Ecuador	203	177	232	213	225	237	207
Equatorial Guinea	19	41	17	4	5	6	12
Gabon	34	42	24	16	10	1	4
Iraq	459	476	341	369	229	419	602
Kuwait	191	303	326	309	204	209	144
Lybia	9	56	43	5	3	12	57
Nigeria	767	406	239	58	54	207	309
Qatar	5	—	—	—	—	—	
Saudi Arabia	1.186	1.361	1.325	1.159	1.052	1.099	943
United Arab Emirates	7		2	13	2	11	20
Venezuela	868	912	755	733	776	741	618
Non-OPEC[a]	4.727	4.495	4.237	4.339	4.690	4.670	4.800
Albania	—	—	—	1	5	7	
Argentina	28	21	13	29	18	12	
Australia	9	6	1	2	10	4	2
Azerbaijan	36	24	29	23	13	10	7
Belize	3	3	2	1	1	—	
Bolivia	6	3	1	—	—	1	
Brazil	232	189	110	145	190	145	198
Brunei	—	2	—	—	—	—	1
Cameroon	36	31	0	—	—	—	
Canada	2.225	2.425	2.579	2.882	3.169	3.227	3.421
Chad	49	28	66	61	72	67	29
China	2	1	1	—	—	—	
Colombia	397	403	367	294	373	442	333
Congo Brazzaville	53	29	18	4	9	3	4
Congo Kinshasa	11	0	0	—	—	—	
Denmark	—	—	—	—	—	2	3
Egypt	4	31	4	—	1	10	8
Ghana	8	—	3	—	—	—	16
Guatemala	9	11	7	7	8	7	8
India	—	—	—	—	2	—	
Indonesia	20	6	18	20	36	34	18
Italy	—	1	—	—	1	—	

Continued

Table 2.3 Evolution of the Imports of Crude Oil by the US During the Period 2011–17 (thousand barrels per day).—cont'd

	2011	2012	2013	2014	2015	2016	2017
Ivory Coast	4	4	—	—	—	—	2
Kazakhstan	7	—	—	—	—	—	
Malaysia	—	—	—	—	0	5	1
Mauritania	—	—	3	2	—	1	3
Mexico	1.102	975	850	781	688	582	608
Netherlands	—	—	1	1	—	1	
Norway	53	26	17	9	9	35	37
Oman	41	9	3	—	—	30	14
Peru	11	8	11	9	6	3	2
Russia	223	101	42	18	38	38	49
Syria	3	—	—	—	—	—	
Thailand	17	20	9	6	—	—	3
Trinidad and Tobago	33	27	8	5	7	9	8
United Kingdom	36	18	21	10	11	19	24
Vietnam	10	10	13	10	9	4	2
Yemen	6	—	—	—	—	—	2

[a]Countries listed under OPEC and non-OPEC are based on current affiliations. OPEC and non-OPEC totals are based on associations for the stated period which may differ from current affiliations. Release Date: 10/31/2017.
Source: EIA database (2018).

What are the reasons for these changes? First, domestic supplies have increased due to the use of a new drilling technique, which involves injecting more than one million gallons of high-pressure water into wells drilled to thousands of feet below the surface. The pressure causes the rock layer to crack so that crude oil (unrefined) flows upward. Because hydraulic fracturing freed oil that was previously inaccessible, the crude oil production in the US grew significantly, especially in Texas, North Dakota, and Alaska (Dhillon, 2014).

Second, crude oil imports by the US in the coming years is expected to remain without significant changes because high gasoline prices, production of more fuel-efficient, and the recession of 2008 led to lower national oil consumption, which fell from 20,8 million barrels per day in 2005 to 18,64 million barrels per day in 2013, or a reduction of 10,4% (Dhillon, 2014).

Although crude oil consumption has not recovered to pre-2005 levels, it started to pick up in 2012, and the EIA predicts that crude oil consumption will continue to rise along with domestic production in the coming years, mainly in the transport sector. However, the participation of crude oil in the electricity generation sector in the US will continue to be very low during the coming years (around 0,5% of the total).

Despite the increase in domestic crude oil production and lower crude oil consumption, the US remains the largest importer of crude oil in the world and spent US$427 billion on crude oil imports in 2013 and beyond. According to EIA sources, it is predictable that the US will continue to be a net importer of crude oil until the year 2035. It has been estimated that the average imports of crude oil by the US will be approximately 30.000 barrels per day between 2015 and 2035. Because the US government now allows exports of light crude oil, the heaviest type of crude oil imports are expected to increase to better adjust to the current configuration of the refineries operating in the country, which will imply that net imports of crude oil will vary very little. The heaviest crudes are expected to account for a growing share of US crude oil imports.

In summary, the following can be stated: in 2012, the US imported 8,527 million barrels per day of crude oil; in 2017, the imports of crude oil by the US decreased by 7,3% to 7,912 million barrels per day. The EIA projects that imports of crude oil from the US will be reduced to 6,8 million barrels per day by 2021; this represents a reduction of 14%, before resuming a trend toward the growth of the imports of this type of fuel. However, others project this trend of decreasing imports to continue in the longer term (Canada's Energy Future, 2013. Energy Supply and Demand Projections to 2035, 2013). In the World Energy Outlook 2012 report, a continued decline in imports of crude oil by the US until 2035 is expected, mainly due to the growing domestic crude oil production.

Given the initial stage of tight oil development, the future results of crude oil production and trade in the US are unusually uncertain at this time. For example, currently Canada's crude oil production is exported entirely to the US, so the increase in the national production of crude oil in this country could pose a risk to the growth of the traditional export market for crude oil from Canada (Canada's Energy Future, 2013-Energy Supply and Demand Projections to 2035, 2013).

Crude Oil Exports by the United States

US exports of crude oil have increased significantly before the lifting of export restrictions on crude oil approved by the US government in December 2015. These exports were mainly to Canada, which was excluded from pre-export limits on crude oil from the US. During the period 2000—13, US crude oil exports rarely exceeded 100.000 barrels per day. By 2015, the US was exporting a total of 465.000 barrels per day, of which a total of 422.000 barrels per day was shipped to Canada (90,7%), a total of 26.000 barrels per day to five other countries (5,6%), and 17.000 barrels per day to the rest of the world (3,7%). The number of countries receiving crude oil exported from the US has increased since the restrictions on exporting US crude oil were removed in December 2015. US crude oil exports have occurred despite relatively small price spreads between international crude oil and domestic crude oil, as well as other factors that should reduce crude oil exports such as falling US crude oil production and added cargo export costs (US Crude Oil Exports are Increasing and Reaching More Destinations, 2016).

Based on the latest publicly available data, US crude oil exports averaged 501.000 barrels per day in the first five months of 2016; this means a total of 36.000 barrels per day or 7,7% more than the daily average of the whole year 2015 (see Fig. 2.13). In recent years, exports of crude oil to destinations other than Canada were often reexported volumes of foreign crude oil or shipments of Alaskan crude oil, which were exempt from export crude oil restrictions. In 2016, US crude oil has been exported to 16 different countries totaling 501 million barrels per day (US Crude Oil Exports are Increasing and Reaching More Destinations, 2016)

In addition to Canada, the destination of the largest and most consistent crude oil export in the US during the first five months of 2016 has been Curacao, followed by The Netherlands. Exports of crude oil from the US to Curacao averaged 54.000 barrels per day until May of that year and to The Netherlands 39.000 barrels per day. Petróleos de Venezuela (PDVSA), Venezuela's state oil company, operates 330.000 barrels a day at a refinery

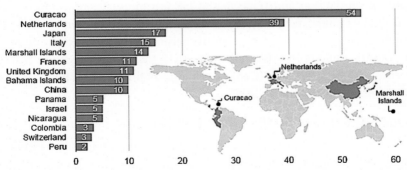

Figure 2.13 US crude oil exports to countries other than Canada. *(Source: EIA database (2016).)*

in Curacao, as well as oil and crude storage facilities on the island. Press reports indicate that US exports of crude oil to Curacao are likely to be used as a diluent, where light (less dense) crude oil is mixed with Venezuelan heavy crude oil, either for processing at the island's refinery or re-export to other PDVSA customers.

It is predictable that continued growth in US crude oil exports will likely depend on increases in US crude oil production and significantly broader price differences between domestic and international crude oil prices. Given the broader climate in the oil market, there are no great incentives or competitive advantages for exporting crude oil from the US. Also, it is important to highlight that foreign refiners have not processed US crude oil for decades, and, for this reason, need to test the processing of US crude oil carefully (Fig. 2.14).

The evolution of US crude oil exports during the period 2012–17 is shown in Fig. 2.15. According to this figure, the US crude oil exports

Figure 2.14 Anacortes refinery. *(Source: Courtesy Walter Siegmund Wikimedia Commons.)*

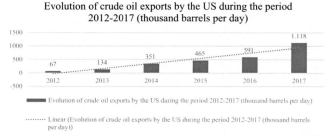

Figure 2.15 Evolution of the US crude oil exports during the period 2012–17 (thousand barrels per day). *(Source: EIA database (2018).)*

increased significantly during the period considered (almost 17 times) from 67.000 barrels per day in 2012 to 1.118.000 barrels per day in 2017. It is expected that US crude oil exports will continue to grow during the coming years.

It is important to stress that when crude oil exports are restricted, US crude oil has problems for its shipping, which causes that this situation is reflected in its prices. In the crude oil export sector, the US and international crude oil are in direct competition, and this makes EIA prices move closer to comparable global crude oil prices. Also, the price of Brent crude oil falls when exports of crude oil from the US are allowed, because as the production of crude oil in this country increases, the supply of global crude oil increases as well.

The Use of Crude Oil for Electricity Generation in the United States

According to EIA sources, in 2017, about 4.015 billion kWh of electricity were generated at utility-scale facilities in the US,[14] a decrease of 1,6% respect to 2016 (4.080 billion kWh). About 63% of this electricity production was generated using fossil fuels (coal, natural gas, petroleum, and other gases) as fuel, about 20% was made using nuclear energy as fuel, and about 17% was created using different renewable energy sources. The EIA estimates that an additional 0,24 trillion kWh of electricity generation and heating were from small-scale solar photovoltaic systems in 2017.

The oil-fired power plants generated just over 0,5% of the nation's electricity in 2017. Due to the negative impact on the environment, petroleum is no longer a favorite source for electricity generation and heating in the US, except Hawaii that gets two-thirds of its electricity from oil. After the growth of OPEC and the oil shocks and price increases of the 1970s, US utilities switched to other fuels, mostly coal.

According to the Electric Power Annual 2015 report, the net electricity generation in the US using petroleum liquids and petroleum coke as fuel in 2015 in all sectors reached the amount of 17.372.000 MWh (5% lower respect to 2014) and 10.877.000 MWh (9,1% lower with respect to 2014),

---

[14] According to EIA sources, there are 1.076 oil-powered electric plants in the US in 2017.

respectively. Per sectors, the electricity generation using petroleum liquids and petroleum coke as fuel in 2015 was the following:

- electricity utilities: petroleum liquids 10.386.000 MWh and petroleum coke 8.278.000 MWh;
- independent power producers: petroleum liquids 6.240.000 MWh (8,1% lower with respect to 2014), and petroleum coke 1.601.000 MWh (13,5% higher with respect to 2014);
- commercial: petroleum liquids 183.000 MWh (26% lower with respect to 2014), and petroleum coke 8.000 MWh (11,2% lower with respect to 2014);
- industrial: petroleum liquids 563.000 MWh (3,5% higher with respect to 2014), and petroleum coke 990.000 MWh (28,8% lower with respect to 2014) (Electric Power Annual, 2015 report).

In 2017, the total electricity generated by oil–fired power plants in the US using petroleum liquids and petroleum cokes as fuel is shown in Table 2.4.

The evolution of the net electricity generation using petroleum liquids and petroleum coke as fuel in the US during the period 2014–17 is shown in Fig. 2.16.

According to the data included in Fig. 2.16, the following can be stated: the net electricity generation using petroleum liquids as fuel in the US

Table 2.4 US Electricity Generation by Source, Amount, and Share of the Total in 2017.

Type of Petroleum	Billion kWh	The Percentage of the Total (%)
Petroleum liquids	13	0,3
Petroleum coke	9	0,2
Total	22	0,5

Source: EIA database (2018).

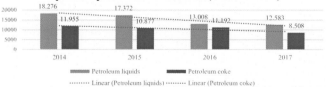

Evolution of the net electricity generation using petroleum liquids and petroleum coke in the US (thousand MWh)

Figure 2.16 Evolution of the net electricity generation using petroleum liquids and petroleum coke in the US during the period 2014–17. *(Source: Electricity Data Browser (2018).)*

during the period 2014—17 decreased by 31,2% falling from 18.276 thousand MWh in 2014 to 12.583 thousand MWh in 2017. It is projected that this trend will continue without change during the coming years. On the other hand, the net electricity generation using petroleum coke as fuel in the US during the same period decreased by 29% falling from 11.955 thousand MWh in 2014 to 8.508 thousand MWh in 2017. It is also expected that this trend will continue without change during the coming years. In both cases, the tendency is to reduce the participation of petroleum in the energy mix of the US in the long-term.

In 2016, coal was the primary source of energy in the electricity generation sector in the US with 33,2% of the total electricity produced in the country in that year (1,35 trillion kWh). Natural gas was the second most used energy source for electricity generation and heating in the US in the same year with 32,7% (1,33 trillion kWh); nuclear energy was the third with 19,5%. In 2017, the situation changed, and natural gas became the primary energy source used to generate electricity with 31,7%, followed by coal with 30,1%, and nuclear energy with 20%. It is projected that natural gas will continue to be the primary energy source used for electricity generation and heating in the US during the coming years. At the same time, it is likely that coal will reduce its role in the energy mix of the country in the future.

Primary energy sources and percent shares of US electricity generation at utility-scale facilities in 2016 and 2017 are listed in Table 2.5. As can be

Table 2.5 Participation of the Different Energy Sources in the US Electricity Generation Sector in 2016 and 2017.

Energy Source	Percentage (%) 2016	Percentage (%) 2017
Coal	33,2	30,1
Natural gas	32,7	31,7
Nuclear	19,5	20
Hydropower	6,1	7,5
Wind	5,6	6,3
Biomass	1,6	1,6
Solar	0,6	1,3
Geothermal	0,4	0,4
Petroleum	0,7	0,5
Other gases	0,3	0,4
Other non-renewable sources	0,3	0,3
Pumped storage hydroelectricity	−0,2[a]	−0,2

[a]Pumped-storage hydroelectricity generation is negative because most pumped storage electricity generation facilities use more electricity than they produce on an annual basis.
Source: EIA database (2018).

seen in the mentioned table, the role of renewable in the energy mix of the country is minimal today, and this situation is not expected to change significantly at least during the coming years. However, in the long-term, the use of renewable for electricity generation and heating will be much higher than today.

Currently, oil covers 36% of the US energy demand, with 71% directed to fuels used in transportation—gasoline, diesel, and jet fuel. Another 24% is used in the industry and manufacturing sectors, almost 5% in the commercial and residential sectors and 0,5% to generate electricity and heating, according to EIA sources. This situation is expected that will not change significantly in the near future, and crude oil will continue to be used for electricity generation and heating but at a very low percentage.

The evolution of the consumption of petroleum liquids and coke for electricity generation and heating in the US during the period 2010—17 is included in Table 2.6.

According to Table 2.6, the consumption of petroleum liquids and petroleum coke for the electricity generation and heating in the US within the period 2010—17 decreased by 45,4% and 33%, respectively. It is expected that this trend will continue without change during the coming years (see Figs. 2.17 and 2.18). For this reason, it can be stated that the role of petroleum liquids and petroleum coke in the US energy mix is projected to decline significantly during the coming years.

Corresponding to Fig. 2.17, the tendency is to systematically reduce the use of petroleum liquids for electricity generation and heating in the US during the coming years, and its role in the energy mix of the country.

Table 2.6 Evolution of the Consumption of Petroleum Liquids and Coke for Electricity Generation and Heating in the US During the Period 2010—17.

Year	Consumption of Petroleum Liquids (thousand barrels)	Consumption of Petroleum Coke (thousand tons)
2010	40.103	4.994
2011	27.326	5.012
2012	22.604	3.675
2013	23.231	4.852
2014	31.351	4.412
2015	28.925	4.044
2016	22.405	4.253
2017	21.935	3.349

Source: EIA Electric Power Monthly (2018).

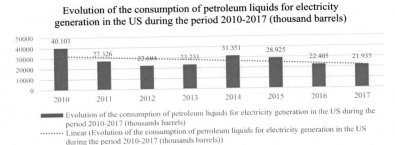

Evolution of the consumption of petroleum liquids for electricity generation in the US during the period 2010-2017 (thousand barrels)

Figure 2.17 Evolution of the consumption of petroleum coke for the electricity generation and heating in the US during the period 2010—17. *(Source: EIA Electric Power Monthly (2018).)*

Evolution of the consumption of petroleum coke for electricity generation in the US during the period 2010-2017 (thousand tons)

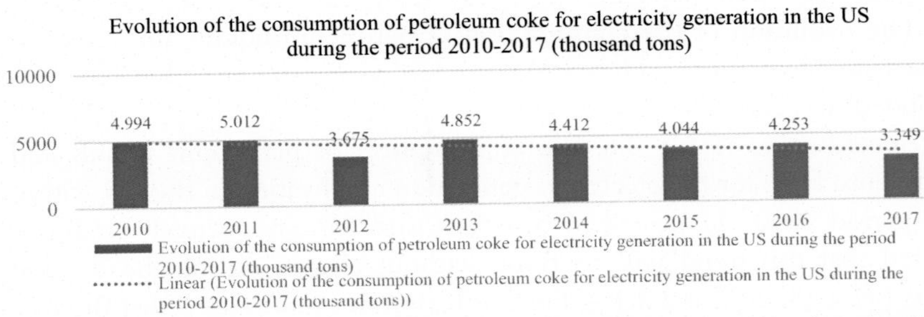

Figure 2.18 Evolution of the consumption of petroleum coke for the electricity generation and heating in the US during the period 2010—17. *(Source: EIA Electric Power Monthly (2018).)*

Within the period 2011—17, the peak in the consumption of petroleum liquids for electricity generation and heating in the US was registered in 2014. If the whole period is considered, then the height in the use of petroleum liquids for electricity generation and heating was recorded in 2010. The primary user of petroleum liquids in the US is the electric power sector. In 2017, the electrical industry used up to 71,7% of the total petroleum liquids consumed by the country in that year. The commercial sector was the sector with the lower consumption of petroleum liquids registered in the US in 2017.

From Fig. 2.18, the following can be stated: it is likely that the use of petroleum coke for electricity generation and heating in the US will be reduced during the coming years, as well as its role in the energy mix of the country. Within the period considered, the peak in the consumption of petroleum coke for electricity generation and heating in the US was registered in 2011. The leading consumer of this type of energy sources is the electric power sector itself. In 2017, this sector consumed 81,5% of the total. The commercial sector is the area with the lowest consumption of petroleum coke for electricity generation and heating during the whole period. In 2017, this sector consumed only 0.09% of the total.

On the other hand, due to the negative impact on the environment and the population, the number of oil-fired power plants used in the US for electricity generation and heating during the period 2010—16 has been reduced significantly (8%) from 1.169 power plants in 2010 to 1.076 in 2016. It is forecast that more oil-fired power plants will be closed in the US during the coming years.

The evolution in the number of oil-fired power plants operating in the country within the period considered is shown in Fig. 2.19. On December 2016, there were around 8.084 power plants in the US of at least 1 MW, and only 13,3% of them are oil-fired power plants (1.075 power plants).

It is important to highlight that the number of final customers of electricity in the US is increasing every year. The evolution in the number of electricity consumers in the US during the period 2010—16 is shown in Fig. 2.19. Based on the data included in the mentioned figure, the following can be stated: the number of consumers of the electricity generated in the US during the period 2010—16 increased by 4,1% rising from 144.140.259 million consumers in 2010 to 150.055.258 million consumers in 2016. It is anticipated that the number of electricity consumers in the US will continue to increase during the coming years.

On the other hand, and according to Fig. 2.20, the number of oil-fired power plants used for electricity generation and heating in the US decreased by 5,1% during the period considered, falling from 1.133 oil-fired power plants in 2005 to 1.076 oil-fired power plants in 2016. The peak in the number of oil-fired power plants used in the US for electricity generation and heating within the period 2005—16 was reached in 2008 (1.170 oil-fired power plants). Since that year, the number of oil-fired power

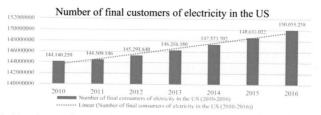

Figure 2.19 Number of electricity consumers in the US during the period 2010—16. *(Source: EIA Electricity Power Annual 2016 (2018).)*

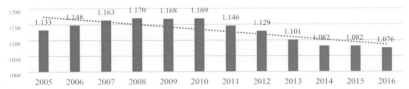

Evolution in the number of oil-fired power plants in the US during the period 2005-2016

Figure 2.20 Evolution in the number of oil-fired power plants used in the US for electricity generation and heating during the period 2005—16. *(Source: EIA Electricity Power Annual 2016 (2018).)*

plants used in the country for electricity generation and heating decreased almost every year until 2016. It is expected that this trend will continue during the coming years as a result of the reduction in the participation of oil in the energy mix of the country. Thirty generators with a total capacity of 1.050,9 MW will be retired during the period 2016—20.

Crude Oil Pipelines in the United States

More than 305.775 km of liquid petroleum pipelines cross the US. These pipelines connect the crude oil extraction areas to the different refineries that operate within the country. Pipelines are safe, efficient, and as most are buried, they are practically invisible. These pipelines transport crude oil from oil fields located on land and offshore to refineries where it is transformed into fuels and other oil products. These products are in turn transported from the refineries to the terminals where they are carried in trucks to the points of sale. The pipelines operate 24 hours a day, 7 days a week.

The map with all major oil petroleum pipelines inside the US is shown in Fig. 2.21.

The Keystone Oil Pipeline

The Keystone Oil Pipeline System is "an oil pipeline system in Canada and the United States, commissioned in 2010 and now owned solely by TransCanada Corporation. It runs from the Western Canadian Sedimentary Basin in Alberta to refineries in Illinois and Texas, and also to oil tank farms and an oil pipeline distribution center in Cushing, Oklahoma" (United States Department of State Bureau of Oceans and International Environmental and Scientific Affairs, 2013).

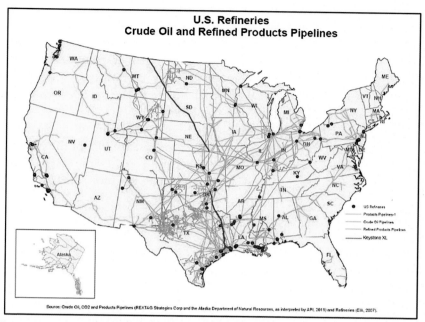

Figure 2.21 Map of the key petroleum mapping inside the US. *(Source: Crude Oil, CO₂, and Products Pipelines (REXTAG Strategies Corp. and the Alaska Department of Natural Resources); EIA; 2007.)*

After six years of revision, former president Obama announced on November 6, 2015, the rejection of his administration to the beginning of the fourth phase of this project. However, on January 24, 2017, president Trump signed presidential memoranda to revive the Keystone XL and Dakota Access pipeline's plans, and to allow the environmental review process to begin (Mufson, 2017).

The Crude Oil Sector in Canada

It is important to stress that the Canadian oil industry developed in parallel with that one of the US and is, and will continue to be for the foreseeable future, the largest foreign supplier of crude oil to the US. The Canadian oil and natural gas industry comprise several hundred companies[15] that carry

---

[15] In 2014, a total of 435 Canadian energy companies were identified as having energy assets either in Canada or abroad. According to Energy Fact Book 2016–17, a total of 59 companies (13% of the total) had energy assets with a value in excess of US$1 billion; a total of 214 companies (49% of the total) had interests in 75 countries; and a total of 166 companies (38% of the total) had energy assets in at least two countries.

out activities such as exploration, drilling, production, and processing of oil and gas in the field, but less than 20 of these companies, however, are responsible for the majority of petroleum production, refining, and marketing. The five largest Canadian oil companies are responsible for more than 50% of oil production in the country. Several firms provide support services for crude oil and natural gas extraction operations, such as contracts for drilling and maintenance of the wells built.

According to Energy Fact Book 2016—17, Canadian oil and gas industries accounted for US$65 billion, which represents 26% of the country capital expenditure registered during the period 2014—15. Oil and gas companies spent additional US$5,2 billion on exploration only in 2014. During the period 2014—15, there were 186 oil and gas deals, which represents 75% of the total sales reported in that period and valued at US$1,9 billion or 88% of the total value of these deals. Canadian energy assets grew, in 2014, to US$543,9, an increase of 12% from the level reached in 2013. Canadian energy assets abroad totaled US$149,7 billion in 2014, an increase of 28% over the 2013 value of US$116,9 billion". In 2014, the majority of Canadian energy assets abroad value were located in the US (70%).

According to Today in Energy (May 2018), Canada crude oil exports to the US reached 3,3 million barrels per day in 2016. In that year, Canada provided 41% of total US crude oil imports. Sales of Canadian crude oil to the US reached more than US$83 billion in 2014. However, as a result of the fall in the oil prices in 2015 and 2016, the value of US crude oil imports from Canada fell from US$47 billion in 2015 to US$36 billion in 2016, a reduction of 23,5%, despite increasing in volume. The decrease is higher (57%) if the year 2014 is taken as reference.

Canada's crude oil exports to the US consist mainly of heavy crude oil because the Canadian oil sands are projected to be the critical diver of growth in the country crude oil production during the coming years.

It is important to note that most of Canada's crude oil and natural gas production takes place in the Sedimentary Basin of Western Canada, which extends from southwest Manitoba to northeastern British Columbia. The basin mentioned above also covers most of Alberta, the southern half of Saskatchewan, and the southwest corner of the Northwest Territories.

The unique geography, geology, resources, and settlement patterns of Canada have been critical factors in the history of the development within the energy sector in Canada. The development of this sector has allowed knowing how this has helped the nation to be entirely different from the US. Unlike the latter country, which has some different oil-producing regions, the vast majority of Canada's oil resources are concentrated in a vast sedimentary basin located in the west of the country, one of the formations with the largest deposits of crude oil in the world. It has an area of 1.400.000 km^2 in western Canada, including most of four western provinces and one northern territory. It consists of a massive wedge of sedimentary rock up to 6-km thick that extends from the Rocky Mountains in the west to the Canadian Shield in the east and is situated far from the ports of the east and west coast of Canada and its major industrial centers. It is also distant from the US industrial centers (Petroleum Industry in Canada, 2018).

"Because to its geographic isolation, the area was settled relatively late in the history of Canada, and its true potential was not discovered until after the end of World War II. As a result, Canada built its major manufacturing centers near its historical hydroelectric power sources in Ontario and Quebec, rather than its petroleum resources in Alberta and Saskatchewan. Not knowing about its crude oil potential, Canada began to import the vast majority of its crude oil needs from other countries as it developed into a modern industrial economy" (Petroleum Industry in Canada, 2018).

It is important to highlight that the province of Alberta lies at the center of the Western Canadian Sedimentary Basin and the formation underlies most of the area, which is the largest oil-producing province throughout the country. However, "the potential of Alberta as an oil-producing province long went unrecognized because it was geologically quite different from American oil-producing regions. On the other hand, "the status of Canada as an oil importer from the US suddenly changed in 1947 when the Leduc No. 1 well was drilled a short distance south of Edmonton. Geologists realized that they had completely misunderstood the geology of Alberta, and the highly prolific Leduc oil field, which has since produced over 50.000.000 m^3 (310.000.000 barrels of crude oil was not a unique formation). There were hundreds of more Devonian reef formations like it underneath Alberta, many of them full of crude oil" (Petroleum Industry in Canada, 2018).

Most of the crude oil companies exploring for oil in Alberta province were US companies. In 1973, when the number of US companies exploring the area reached its peak, "over 78% of Canadian crude oil and natural gas production and over 90% of oil and gas production companies were under foreign control or ownership, mostly US firms. This foreign ownership spurred the National Energy Program under the Trudeau government" (Tertzakian, 2012).

Finally, it is important to single out that the "Canadian Regulatory Authority rests on provincial governments. Federal authority over interprovincial and international oil and gas pipelines rests on the National Energy Board. The Canada—Newfoundland and Labrador Offshore Petroleum Board and Canada—Nova Scotia Offshore Petroleum Board are responsible for the regulation of crude oil and gas activities in their corresponding offshore areas" (Energy Markets Fact Book-2014/2015, 2014).

Crude Oil Reserves in Canada

It is well-known today that Canada is one of the most resource-rich nations on the planet. According to Energy Fact Book 2016—17, the country's proved crude oil reserves are vast enough to meet its energy demands for 120 years at the current rate of production of 3,9 million barrels per day. Considering its level of proved crude oil reserves of 171 billion barrels, Canada occupied, in 2017, the third position at world level, with the 10,3% of the world total proved crude oil reserves.

Much of Canada's proved crude oil reserves consist of oil contained in the oil sands of Alberta (estimated in 2016 at 165,3 billion barrels or 97% of the total crude oil reserves of the country). The oil sands are located in three central regions within the province of Alberta. These regions are Athabasca, Cold Lake, and Peace River, which combined cover an area more than 142.000 km^2. However, other non-oil sands deposits are very popular across all of western Canada in what is known as the "Western Canadian Sedimentary Basin." According to December 2016 report in the Oil and Gas Journal, "Canada's proved crude oil reserves raised to 180 billion barrels thanks to inclusion of the oil sands - also known as tar sands - now considered recoverable with existing technology and market conditions" (Walsh, 2017); this means an increase of 5,9% in the level of the crude oil reserves of the country since 2016.

However, extracting crude oil from the vast majority of Canada's crude oil reserves is a labor and capital-intensive process, because production tends to come in sporadic bursts rather than steady streams. Oil companies, therefore, begin by extracting lower density, higher-value crude oil first, and directing their efforts in removing crude oil from other types of deposits only in times of high commodity prices.

The evolution of the crude oil reserves in Canada during the period 2010—17 is shown in Fig. 2.22.

According to Fig. 2.22, the proved crude oil reserves in Canada decreased by 1,8% during the period 2010—15 falling from 175 billion barrels in 2010

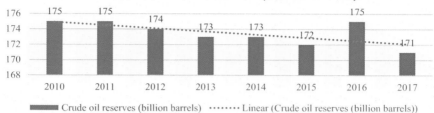

Figure 2.22 Evolution of the proved crude oil reserves in Canada during the period 2010—17. *(Source: EIA database (2017) and National Resources Canada (2017). According to Oil and Gas Journal, Canadian proved crude oil reserves reached 180 billion barrels in 2016.)*

to 172 billion barrels in 2015. However, during the period 2015—16, proved crude oil reserves increased by 1,8% rising from 172 billion barrels in 2015 to 175 billion barrels in 2016, for the first time since 2010. During the period 2016—17, Canadian proved crude oil reserves decreased again by 2,3% falling from 175 billion barrels in 2016 to 171 billion barrels in 2017. Canada's proved crude oil reserves declined by 2,3% within the period 2010—17. It is expected that the current trends in the level of the country crude oil reserves will continue without change during the coming years unless new crude oil deposits are found within the country.

It is important to single out that Canada's proved crude oil reserves during the period from 1980 to 2002 "were well below 10 billion barrels. In 2003, they rose to 180 billion barrels after oil sands resources were deemed to be technically and economically recoverable (Western Canadian Oil and Gas, Exploration and Production Industry, 2014). Oil sands now account for 165,3 billion barrels of proved crude oil reserves, 96,5% of all of Canada's current proved crude oil reserves. Oil sands proved reserves saw a slight year-on-year decline during the period 2010—15. Nevertheless, oil sands production continued to be robust, despite the decrease in crude oil prices (Canada Oil Market Overview, 2016). The increase in oil sands proved reserves reported in 2016 was thanks to the inclusion of the crude oil sands reserves now considered recoverable with existing technology and market conditions.

Crude Oil Production in Canada

Crude oil is produced in Canada across the country from coast to coast. In 2017, Canada was considered as the fifth largest global oil producer (after

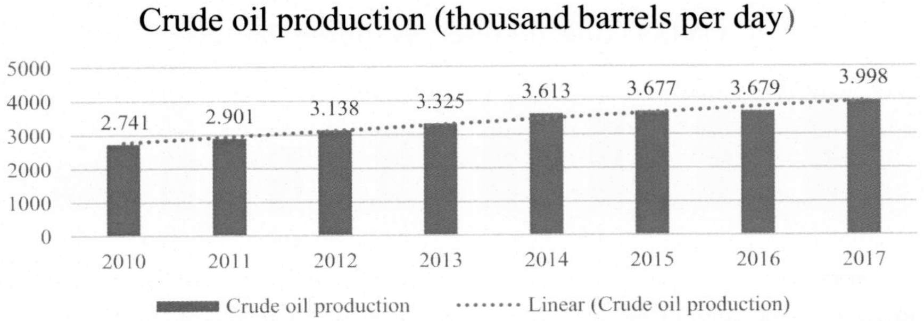

Crude oil production (thousand barrels per day)

Figure 2.23 Evolution of crude oil production in Canada during the period 2010—17. *(Source: EIA database (2018).)*

the US, Russia, Saudi Arabia, and Iran), extracting 3.998 thousand barrels per day according to EIA database (2018). In that year Alberta had the highest amount of crude oil production in Canada. It is projected that, in 2040, crude oil production in Canada could reach 6,3 million barrels per day, an increase of 61,5% respect the crude oil production reported in 2016.

Crude oil production in Canada comes from three principal sources: (1) the oil sands; (2) the resources in the broader Western Canada Sedimentary Basin; and (3) the offshore oil fields in the Atlantic Ocean (Crude Oil: Forecast, Markets, and Transportation, 2017).

The evolution of the crude oil production in Canada during the period 2010—17 is shown in Fig. 2.23.

According to Fig. 2.23, the crude oil production in Canada during the period 2010—17 increased by 45,9% rising from 2.741.000 barrels in 2010 to 3.998.000 barrels in 2017. According to Crude Oil: Forecast, Markets, and Transportation (2017) report, crude oil production is expected to reach 4,62 million barrels per day in 2020, an increase of 16% respect to 2017; 4,88 million barrels per day in 2025, an increase of 5,6% respect to 2020; 5,12 million barrels per day in 2030, an increase of 4,9% respect to 2025; and 6,3 million barrels per day in 2040, an increase of 23% respect to 2030.

Canada's crude oil production currently meets close to 40% of the country total energy needs through a variety of products created by refining its crude oil production. On average, and according to Crude Oil and Canada (2016) report, refining crude oil yields the following range of products:

- Gasoline to fuel cars, trucks, aircraft, and emergency generators, among other equipment;
- Diesel fuel to power cars, trucks, buses, railway locomotives, boats and ships, and more significant electric generators;

- Asphalt for road paving and roofing, lubricants, waxes for candles and polishes, and several materials for the petrochemical industry;
- Heavy fuel oil for electric power generation, large ships, and some industrial processes;
- Light fuel oil for heating homes and buildings, many industrial processes, and the fuel for some vessel;
- Aviation jet fuel for turbine-powered airplanes;
- Propane and butane;
- Kerosene, greases, petroleum coke (Crude Oil and Canada, 2016).

Light Tight Oil Reserves and Production in Canada

Over 500,000 oil and natural gas wells have been drilled in Canada since the first commercial oil well in the region was dug in Oil Springs, Ontario, in 1858 (Exploration and Production of Shale and Tight Resources, 2016). In Canada, tight crude oil production activities were initiated during the period 2005—06, and are located primarily in western Canada. Exploration has been pursued in the country in only a minimal number of provinces.

Light tight oil is found in sedimentary rock characterized by very low permeability, typically shale rock, and are extracted by using horizontal drilling combined with multi-stage hydraulic fracturing applied to low permeability light crude oil reservoirs (Crude Oil Facts, 2018). There are two main types of tight oil:

- Oil found in the original shale source-rock, similar to shale gas and typically called "shale oil";
- Oil that has migrated out of the original shale source rock into nearby or distant tight sandstones, siltstones, limestones, or dolostones (Tight Oil Developments in the Western Canada Sedimentary Basin, 2011).

According to Energy Fact Book 2016—17, in 2015, the world technically recoverable shale oil reserves reached 419 billion barrels. Canada shale oil reserves represent 2% of the country total crude oil production in that year (8,38 billion barrels). In December 2014, the output of tight oil in Canada had grown to 450.000 barrels per day, doubling from early 2011, and accounted for more than 10% of total Canadian crude oil production. However, in January 2016, the output of tight oil in Canada dropped to 360.000 barrels per day; this represents a reduction of 20% respect to December 2014. It is important to highlight that according to the Canada National Energy Board, the growth rate of tight oil production in the country has been slowing in the last years. In March 2012, production of light tight oil was 127.000 barrels per day higher than the level reached in March 2011. By April 2014, this growth slowed to 33.000 barrels per day, a reduction of 74% respect to the level achieved in March 2012.

Considering the International Energy Outlook 2016 report, tight oil production in Canada is expected to continue to decline until 2020, and then it is forecast to increase over the rest of the projection period until 2040, reaching 760.0000 barrels per day in that year. This outcome is in response to higher crude oil prices and less competition for capital with oil sands development.

By 2030, it is expected "moderate growth in light oil production from tight oil plays in the Western Canadian Sedimentary Basin as conventional heavy oil production declines. However, the development of tight oil reservoirs is still in early stages in Canada. The extent to which these resources can be produced from is largely undetermined" (Exploration and Production of Shale and Tight Resources, 2016).

Oil Sands Reserves and Production in Canada

According to Oil and Gas Journal and the Canadian Association of Petroleum Producers, Canada oil sands reserves are estimated, in 2017, in 165,3 billion barrels, which represents 96,5% of the total crude oil reserves existing in the country.

Production from the oil sands accounted for more than half of Canadian crude oil and other liquid's output in 2016 (62% of the total), a proportion that has steadily increased over the past decade. According to Crude Oil Facts (2018), oil sands production has generally been growing since 2006, peaking at 2,7 million barrels per day in 2017 compared to 2,42 million barrels per day in 2016,[16] an increase of 11,6%, and has an estimated US$288 billion of capital investment to date, including US$16,6 billion in 2016. Conventional crude oil production has remained stable at about 1,5 million barrels per day with a peak in 2014. Considering all available data, the output of oil sands in Canada is likely to continue to grow after 2017, while the production of conventional crude oil is likely to decrease during the same period. In 2040, it is probable that oil sands

---

[16] Capital investment in the oil sands is expected to reach US$15 billion during the period 2017—18, which is 56% lower than the US$34 billion capital investment reported in 2014 (Crude Oil: Forecast, Markets, and Transportation, 2017).

production will represent 71% of the total crude oil production in the country, an increase of 9% since 2016.

About 81% of Alberta's total crude oil production came from oil sands (Alberta's Energy Reserves and Supply/Demand Outlook, 2016). "Other noteworthy producing provinces are Saskatchewan, with roughly 15% of national output from its share of the Western Canada Sedimentary Basin, and areas off the east coast of Canada, primarily offshore Newfoundland, and Labrador" (Statistical Handbook for Canada's Upstream Petroleum Industry, 2015). "Because production from offshore reserves comes from mature oilfields with typically declining production rates, western provinces are expected to comprise an increasing proportion of overall Canadian crude oil and other liquid's production in the coming years" (Crude Oil: Forecast, Markets, and Transportation, 2017).

The oil sands production in Canada has been conducted commercially for almost five decades reaching 2,42 million barrels per day in 2016. The projections for oil sands output during the period 2016—40 are shown in Fig. 2.24.

According to Fig. 2.24, the following can be stated: the production of oil sands in western Canada during the period 2016—40 is expected to increase by 86% from 2,42 million barrels per day in 2016 to 4,5 million barrels per day in 2040. It is important to single out that oil sands resources (three deposits) are situated almost entirely in the Alberta region and must be extracted by drilling. With the development of new and more efficient extraction methods, bitumen and economical synthetic crude oil are produced at costs nearing that of conventional crude oil.

It is important to know that initially, oil sands were mainly accessed through large open pit mining operations. Since the mid-1980s and

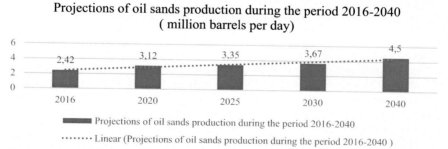

Projections of oil sands production during the period 2016-2040
(million barrels per day)

Figure 2.24 Projections of oil sands production in western Canada during the period 2016—40. *(Source: Crude Oil: Forecast, Markets, and Transportation (2017).)*

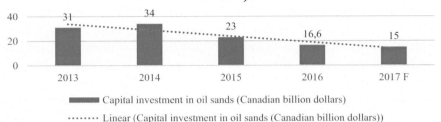

Figure 2.25 Evolution of capital investment in oil sands in Canada during the period 2013—17. *(Source: Crude Oil: Forecast, Markets, and Transportation (2017); Note: F (forecast).)*

especially over the last decade, in situ technologies have played a growing role in oil sands production.[17] In situ oil sand production can be broadly categorized into two types: (1) primary; and (2) thermal. Primary production is heavy crude oil that flows naturally into a well without the use of heat, and it has become a smaller component of the total in situ crude oil production, shrinking from 42% in 2000 to 23% in 2014; this means a reduction of 19%. It is probable that this trend will continue in the coming years and a further decrease of 12% is projected to be registered by 2040.

The evolution of the capital investment in the oil sands sector in Canada during the period 2013—17 is shown in Fig. 2.25.

Analyzing the data included in Fig. 2.25, it can be stated that the capital investment in oil sands during the period 2013—17 has been reduced by 52%. The peak in capital investment in oil sands during the period considered was reached in 2014 when the capital investment grasped 34 billion Canadian dollars. After that peak, the capital investment in oil sands dropped 56%. It is expected that the capital investment in the oil sands sector in Canada will change this trend during the coming years. Since the

---

[17] Recent improvements in productivity are partly due to producers responding to low prices by only drilling their most productive assets. Over time, it is likely that activity will move to less prolific areas of producing basins, which would have a downward impact on productivity. The extent to which technology and improved efficiencies can offset this is an open question. The projections made by the National Energy Board in its Canada's Energy Future 2017 report, "take into account recent productivity trends, with productivity increasing or decreasing depending on the producing area. Over the long-term, productivity is held constant or declines, depending on the characteristics of particular producing regions".

beginning of the extraction of oil sands in Canada, a total of US$271 billion has been invested in this sector.

Finally, the following can be stated: Canada is uniquely positioned to supply an abundance of safe, secure energy as fossil fuels will satisfy the vast majority of growing global demand — which will increase as economies grow and standards of living improve. Canada is forecasted to be fourth in oil production growth at world level during the period 2015—40 after Iraq, Brazil, and Iran. Steady growth is predicted for Canadian crude oil production to 6,3 million barrels per day by 2040. In the case of oil sands, its output is predicted to grow up to 4,5 million barrels per day in 2040; this represents an increase of 86% respect to 2016.

On the other hand, with crude consumption rising worldwide and conventional crude oil supplies declining, the need for a secure amount of crude oil from unconventional resources like Canada's oil sands will continue to increase. With the majority of Canadian crude oil reserves located in the oil sands deposits, the country has the potential to become a key global supplier of this type of energy source. Oil sands reserves have helped diversify global supply, reduced reliance on more distant sources of crude oil, and improved North American energy security while supporting economic growth in Canada and the US.

Canada's oil and natural gas industry provide economic benefits across the North America region. Almost 1.580 US-based companies are providing goods and services to Canada's oil sands sector, including constructions, electrical, engineering, and equipment services.

Summing up and according to the content of the Crude Oil: Forecast, Markets, and Transportation Report (2017), the following can be stated:

- Western Canada is and will continue to be during the coming years, the primary source of supply and future crude oil production growth in the country. In eastern Canada, overall crude oil production is expected to decline during the coming years;
- Eastern Canada's contribution is expected to increase to 307.000 barrels per day in 2024 but is projected to fall to 186.000 barrels per day by 2030; this means a reduction of 40%;
- Western Canada oil sands production is expected to grow from 2,42 million barrels per day in 2016 to 4,5 million barrels per day in 2040; this means an increase of 86% during the whole period considered;
- Western Canada conventional crude oil production is expected to contribute 884.000 barrels per day on average throughout the projected period;

- Additional 1,5 million barrels per day of crude oil supply from western Canada is forecast to enter the global market by 2030, of which over 90% will be heavy crude oil;
- From 2017 to 2020, the growth in crude oil supply is projected to be, on average, 5% per annum before falling to a slower average annual rate of growth at 2% during the period 2021–30;
- Long-term oil sands production growth in Canada could be affected by (1) indications of possible US protectionist policies; (2) federal and provincial climate change policies, which impose constraints on the ability to reduce capital and operating costs; and (3) divergent environmental policies between Canada and the US (Crude Oil: Forecast, Markets, and Transportation, 2017).

Crude Oil and Other Liquid Fuels Consumption in Canada

According to the Energy Markets Fact Book-2014-15 report, "a total of 12% of Canadian consumption of refined petroleum products is imported, and 74% of Canadian imports come from the US. Another Canadian crude oil or refined petroleum product imports come from a wide range of countries, including The Netherlands (9%), Mexico (3%), and the UK (3%)."

The evolution of the consumption of crude oil in Canada during the period 2010–17 is shown in Fig. 2.26.

According to Fig. 2.26, the consumption of crude oil in Canada increased by 5,3% during the period 2010–17, rising from 2.306 thousand barrels per day in 2010 to 2.428 thousand barrels per day in 2017; the

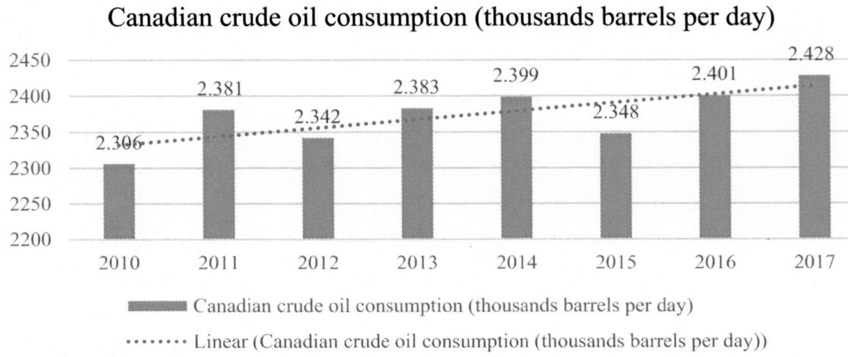

Canadian crude oil consumption (thousands barrels per day)

Figure 2.26 The evolution of the consumption of crude oil in Canada during the period 2010–17. *(Source: BP Statistical Review of World Energy 2018 (June 2018).)*

highest peak of crude oil consumption during the period under consideration was reached in 2017. It is expected that the use of crude oil in Canada will continue to increase at least during the coming years, but not for electricity generation and heating.

It is important to single out that in Canada, fuel efficiency gains result in the relatively flat consumption of petroleum and other liquid fuels. By 2035, the energy used per unit of economic output is projected to be 20% lower than in 2012, due to improvements in energy efficiency (Canada's Energy Future, 2013 – Energy Supply and Demand Projections to 2035, 2013).

Crude Oil Exports by Canada

Canada produces more crude oil than it can consume. As a result, Canada is a significant net exporter of crude oil being the fourth largest exporters of crude oil in the world. According to Crude Oil Facts (2018), 99% of Canada's crude oil exports go to the US.[18] Two-thirds of Canadian crude oil exports are delivered to refineries in the US Midwest. Relatively small amounts of Canadian crude oil are shipped overseas to Europe, Asia, and South America. Having in mind Canada's Energy Future 2016, Energy Supply and Demand Projections to 2040 report, the following can be stated: "the majority of growths in Canadian crude oil exports is likely to be heavy crude oil as the oil sands are projected to be the main driver of growth in Canadian oil production. Also, the US market is likely to be well supplied with light oil due to growing US light oil production".

It is important to note that the growth of Canadian crude oil exports to the US Gulf and West coasts will depend on the increase in crude oil demand and the extent to which Canadian crude oil producers can compete with suppliers of heavy crude oil in the Middle East, Mexico, and South America. There is a variety of potential destinations for Canada's exports of crude oil outside the US. Currently, part of the light crude oil produced off the coast of Newfoundland and Labrador is exported to markets in the Atlantic basin. Depending on the evolution of the crude oil

---

[18] Canada exported 3,08 million barrels per day to the US in 2016, which represents 99% of all Canadian crude oil and equivalent exports in that year, 43% of the total US crude oil imports, and 20% of US refinery crude intake. During the period March—August 2017, Canada exported to the US between 3.768.000 and 4.213.000 barrels per day of crude oil and other products and from 3.241.000 to 3.569.000 barrels per day of crude oil (EIA database).

transport infrastructure in the future, heavy crude oil from western Canada could potentially reach markets in Europe, where there is significant refining capacity for this type of fuel.

On the other hand, the producers of heavy crude oil from western Canada have identified Asia as a potential long-term growth market for the export of its heavy crude oil production, due to the particular characteristics of the type of crude oil. Canada is well positioned geographically to supply to the Asian markets with this type of crude oil since the distance from the west coast of Canada to Asia is shorter compared to the distance between other countries producing heavy crude oil and the mentioned region. However, future developments in crude oil pipeline infrastructure will be an essential factor for Canadian crude oil producers to access markets outside the country.

The evolution of Canada's world export of crude oil and other oil products during the period 2013—17 is shown in Fig. 2.27.

From the data included in Fig. 2.27, the following can be stated: the exports of crude oil from Canada during the period 2013—17 increased by 27% rising from 2.616,18 million barrels per day in 2013 to 3.322,93 million barrels per day in 2017. It is expected that the exports of crude oil from Canada will continue to grow at least during the coming years.

According to Fig. 2.28, it can be stated that the export of crude oil and other oil products from Canada to the US increased by 38,9% during the period 2011—16. During the period considered, there was an annual increase in the export of crude oil and other oil products from Canada to the US. It is expected that this trend will continue at least during the coming years, despite the increase in the production of crude oil in the US reported in recent years.

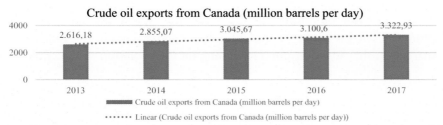

Figure 2.27 Evolution of Canada's world export of crude oil and other oil products during the period 2013—17. (Source: 2017 Crude Oil Annual Export Summary, National Energy Board (2018).)

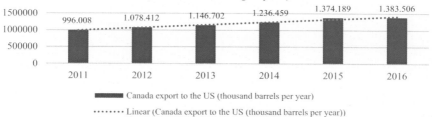

Figure 2.28 Evolution of the export of crude oil and other oil products from Canada to the US during the period 2011−16. *(Source: EIA database (2017).)*

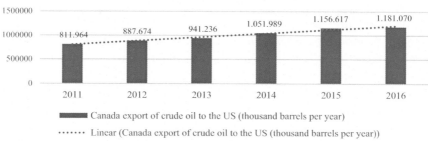

Figure 2.29 The evolution of the export of crude oil from Canada to the US during the period 2011−16. *(Source: EIA database (2017).)*

In the specific case of crude oil, the evolution of the export of this product from Canada to the US during the period 2011−16 is shown in Fig. 2.29.

From the data included in Fig. 2.29, the following can be stated: the export of crude oil from Canada to the US increased by 45,5% during the period 2011−16. During the period considered, there was an annual increase in the shipping of crude oil from Canada to the US. It is expected that this trend will continue without change during the coming years. However, the increase in the exports of crude oil from Canada to the US could be affected as a result of an increase in the production of crude oil in the US, particularly the production of tight and shale oil.

According to Fig. 2.30, Canadian crude oil exports have been increasing since 2010, reaching 3,3 million barrels per day in 2016. Imports of crude oil by Canada reached 0,91 million barrels per day in 2006 but were steadily

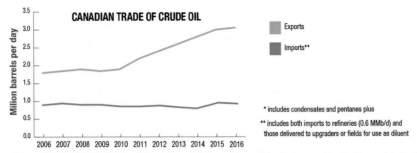

Figure 2.30 Canada's trade of crude oil during the period 2006–16. *(Source: Crude Oil Facts. Natural Resources Canada.)*

declining between 2006 and 2014 (reaching 0,80 million barrels per day in 2014). Since 2015, the imports of crude oil by Canada began increasing again, reaching 0,91 million barrels per day in 2016, the same level of 2006.

The US and Canada share the world's largest and most comprehensive trading relationship. Energy is a significant part of this relationship as Canada is the single largest foreign supplier of power to the US with the oil sands being the most significant source. It is important to highlight that Canada is uniquely positioned to contribute to meeting growth in US energy demand projected for the coming years.

Growing domestic crude oil supply helped Canada overtake the OPEC for cumulative US imports by the end of 2014. The US crude oil exports to Canada were 10 times higher in 2015 than five years ago. Meanwhile, Canadian oil exports to the US have increased by about 40% over the same period. Even with increased domestic oil supply, the US will need to import crude oil from other countries with the purpose to meet its huge energy demand.

Crude Oil Imports by Canada

Due to the regional nature of Canadian refining markets, Canada also imports some crude oil from a wide range of countries located in North America, the Middle East, Africa, and Europe. In 2018, according to Crude Oil Facts (2018) report, these countries were the following: US (61%), Saudi Arabia (12%), Azerbaijan (6%), Norway (5%), and Nigeria (4%) (see Fig. 2.31).

Due to its proximity with the US, the majority of the crude oil that Canada imports each year comes from this country (61%), and this situation is not expected to change during the coming decades.

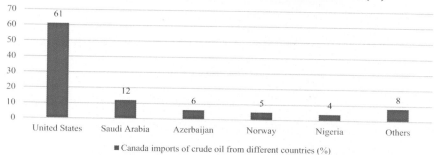

Figure 2.31 Imports of crude oil by Canada and by country in 2018. *(Source: Crude Oil Facts (2018).)*

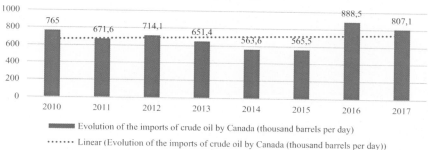

Figure 2.32 Evolution of the imports of crude oil by Canada during the period 2010—17 (thousand barrels per day). *(Source: OPEC database (2018).)*

The evolution of the imports of crude oil by Canada during the period 2010—17 is shown in Fig. 2.32.

According to Fig. 2.32, the imports of crude oil by Canada during the period 2010—17 increased by 5,5% rising from 765 thousand barrels per day of crude oil in 2010 to 807,10 thousand barrels per day of crude oil in 2017. It is expected that the imports of crude oil by Canada will continue to increase at least during the coming years.

Key Oil Pipelines in Canada

Canada has an extensive network of 840.000 km of pipelines carrying crude oil to domestic and US refineries, including 117.000 km of large-diameter transmission lines (Pipelines Across Canada, 2016). The current

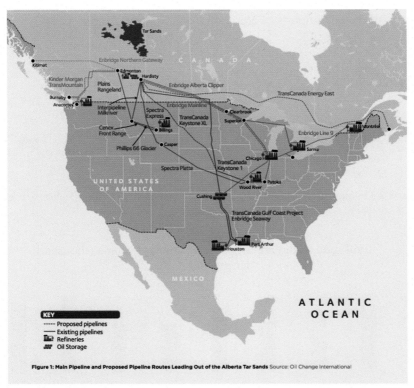

Figure 1: Main Pipeline and Proposed Pipeline Routes Leading Out of the Alberta Tar Sands Source: Oil Change International

Figure 2.33 Main Pipeline and Proposed Pipeline Routes Leading Out of the Alberta Tar Sands. *(Source: Oil Change International, 2015.)*

crude oil pipeline capacity existing in western Canada is estimated at 3,9 million barrels per day[19] (Crude Oil Facts, 2018), and transported, in 2016, a total of 3.583.000 barrels per day. A total of 31 operating oil pipelines cross the border between Canada and the US.

According to Fig. 2.33, and the Energy Fact Book 2016—17, the main components of Canada's crude oil pipelines are:

1. **Enbridge:** It is the world's largest pipeline system for crude oil and petroleum products, serving Canada and the US with an estimated capacity of 2.850.000 barrels per day.

---

[19] By 2030, the annual supply of western Canada crude oil forecast to be 5,4 million barrels per day, an increase of 1,5 million barrels per day from the level registered today (Crude Oil: Forecast, Markets, and Transportation, 2017).

2. **Kinder Morgan:** It is North America's largest pipeline company and the primary transporter of refined oil products within the region.
3. **Pembina:** It is the second-largest oil pipeline system in western Canada.
4. **Portland–Montreal Pipeline:** It transports crude oil from Portland, Maine, to Montreal, Canada.
5. **TransCanada Pipeline:** It can transport 591.000 barrels per day of crude oil from Hardisty, Alta, to the US Midwest.
6. **Trans–Northern Pipeline:** It transports petroleum products from Montreal to eastern Ontario, Toronto, and Oakville.

Although crude oil is typically moved by pipeline, a substantial amount of crude oil is also transported by rail. The cargo of fuel oils and crude oil by rail car loadings almost tripled between 2011 and 2014, but subsequently decreased by 16% from 2014 to 2015, due to low crude oil prices. Between 2015 and 2016, the tonnage of fuel oils and crude oil rail car loadings decreased by another 24%; between 2016 and 2017 the average total load was 131 million barrels per day, an increase of 18% concerning 2016. The new peak in the transport of crude oil by rail was reached in 2018 (196 million barrels per day) (Crude Oil Facts, 2018). The estimated rail loading capacity out of western Canada in 2016 is approximately one million barrels per day (Energy Fact Book 2016–2017, 2017).

It is important to single out that current Canadian crude oil production is nearing the maximum pipeline capacity out of western Canada. With western Canadian crude oil production projected to grow over the coming years, several pipeline projects are being proposed to move new crude oil production to different markets. Below is a list of some of the most significant pipelines proposed to be constructed in Canada with the purpose to transport the foreseeable new crude oil production to different markets:

1. **Northern Gateway (Enbridge):** Two new pipelines are planned to be constructed from Edmonton to Kitimat (British Columbia) with a capacity to export 525.000 barrels per day of crude oil and to import 193.000 barrels per day of condensate. A marine terminal would also be constructed. The mainline from Alberta to the US Midwest and Ontario has an estimated capacity of 2.850.000 barrels per day (Crude Oil Facts, 2018).

2. **Trans Mountain Expansion (Kinder Morgan)**[20]: Twinning of the existing pipeline from Edmonton to Vancouver with an incremental capacity of 590.000 barrels per day. A marine terminal in Burnaby (British Columbia) would also be expanded.
3. **Keystone XL (TransCanada):** A new pipeline from Hardisty to US Gulf Coast is being planned with a capacity of 830.000 barrels per day.
4. **Mainline Expansion (Enbridge):** Expansion of Alberta Clipper from Hardisty to Gretna, with an incremental capacity of 350.000 barrels per day.
5. **Energy East (TransCanada):** Conversion of the existing natural gas pipelines to oil and construction of new oil lines with a capacity of 1,1 million barrels per day (Energy Markets Fact Book-2014-2015, 2014).

Finally, it is important to note that the volume of crude oil exported by Canada to the US by rail increased from almost 16.000 barrels per day in the first quarter of 2012 to about 160.000 barrels per day through 2014, due to current pipeline's constraints. The existing rail loading capacity out of western Canada is approximately 300.000 barrels per day but is forecasted to more than double — to 700.000 barrels per day — by the end of 2018.

Electricity Generation Using Oil as Fuel in Canada

According to Statistics Canada database, in 2015, total installed electricity generation capacity in Canada reached 140,7 GW, which represents 2,6% of the world electricity generating capacity installed in that year. The country ranks seventh regarding the total installed capacity of electricity generation at the world level. In 2016, the total installed electricity generation capacity raised to 143,4 GW, an increase of 1,9% with respect to 2015 (see Fig. 2.34). Within the different energy sources used for electricity generation in the country, fossil fuels were the second most important source of electricity in Canada in 2016. About 1,3% of the total power

---

[20] "Federal approval of the expansions of Kinder Morgan's Trans Mountain pipeline to the Pacific and Enbridge Line three to the Midwest US appear promising, assuming you can stay in business until 2019. In politics pipeline approvals are big news. Lots of handshakes and photographs. But in the real-world oil patch, like the one described herein, what is required is pipe carrying oil and gas leading to higher volumes and netbacks creating increased cash flow for reinvestment. Today, not only are the benefits from these pipelines not happening now, but at least for Kinder Morgan they may never happen at all" (Durden, 2016).

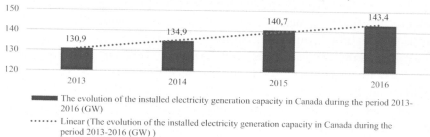

Legend:
The evolution of the installed electricity generation capacity in Canada during the period 2013-2016 (GW)

Linear (The evolution of the installed electricity generation capacity in Canada during the period 2013-2016 (GW))

Figure 2.34 The evolution of the installed electricity generation capacity in Canada during the period 2013—16. *(Source: Statistics Canada (2018).)*

generated by the country is using oil as fuel, which is a percentage very low. However, it is likely that the during the coming years this percentage will be reduced even further.

The evolution of the installed electricity generation capacity in Canada during the period 2013—16 is shown in Fig. 2.34.

According to Fig. 2.34, the evolution of the installed electricity generation capacity in Canada during the period 2013—16 increased by 9,5% rising from 130,9 GW in 2013 to 143,4 GW in 2016. It is expected that the electricity generation capacity in the country will continue this trend during the coming years, despite the adoption of additional measures to reduce, as much as possible, the growth in the consumption of electricity within the country.

Considering all available energy source in the country, the following can be stated: In 2016, hydroelectricity remains the primary source of electric power in Canada (80,8 GW), accounting for 56,3% of total power capacity installed in the country (143,4 GW) (Installed Plants, Annual Generating Capacity by Type of Electricity Generation-Statistics Canada, 2018). Natural gas, coal, and nuclear power plants provide most of the remaining supply (36%), while other renewable energy sources such as wind, solar, and biomass make up 7,7% of the total electricity generation capacity installed. The electricity supply mix varies significantly among the Canadian provinces and territories, reflecting the different types of energy sources available in the country, economic considerations, and policy choices (Canada's Energy Future, 2016. Energy Supply and Demand Projections to 2040, 2017).

By the information included in the report mentioned above, the following can be stated: in 2014, oil-fired power plants accounted for 2,5% of the total Canadian electricity generation installed capacity. These capacities are used to generate electricity during peak demand periods or in areas where other generation options are not widely available, such as Yukon, Northwest Territories, and Nunavut. Total oil-fired capacity is expected to decline by 20,6% during the period 2014−40 falling from 3,4 GW in 2014 to 2 GW in 2040. This reduction reflects the retirements of oil-fired aging units, which are being replaced by renewable power, natural gas, or LNG-fired units when possible.

Due to low utilization rate, oil-fired power plants currently account for roughly half a percent of total generation and maintains a tiny share over the projection period.

The evolution of the electricity generation using oil as fuel in Canada during the period 2010−15 is shown in Fig. 2.35.

Based on the data included in Fig. 2.35, the following can be stated: the electricity generation using oil as fuel in Canada decreased by 20% during the period considered falling from 1,32 TWh in 2010 to 1,06 TWh in 2015. It is predictable that the electricity generation in the country will continue following this trend during the coming years but a slow rate, as a result of different energy measures adopted by the government to reduce the use of crude oil for electricity generation.

Canadian electricity demand, in 2017, was 552 TWh and accounted for 17% of total Canadian end-use energy demand. According to the level of the electricity demand, Canada occupies the sixth place among the 10 top countries at world level. These countries are China, with 5.683 TWh; the

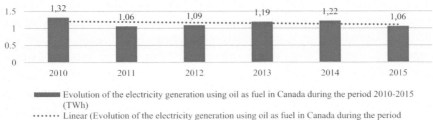

Figure 2.35 Evolution of the electricity generation using oil as fuel in Canada during the period 2010−15. *(Source: The World Bank- World Development Indicators.)*

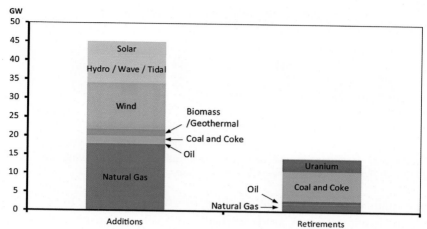

Figure 2.36 Additional and retirements capacities by 2040. *(Source: Canada's Energy Future 2016. Energy Supply and Demand Projections to 2040, 2017.)*

US, with 3.808 TWh; India, with 1.156 TWh; Japan with 1.019 TWh; and Russia with 889 TWh (Global Energy Statistical Yearbook, 2018).

Most of the growth in energy demand is expected to come from the industrial sector, where the overall request is projected to be 0,7%. Total electricity generation capacity is projected to grow by an annual average of 1% during the projection period, reaching 173 GW in 2040 and will cover the retiring units and meet growing demand. The majority of additions to capacity (45 GW) are in the natural gas sector, followed by wind and hydropower facilities, accounting for 84% of the total additions projected for the period 2014—40 (see Fig. 2.36). The remaining additions (16 GW) include 3,5 GW of solar, 1,8 GW of biomass/geothermal, 2 GW of coal, and 8,7% of crude oil. Retirements are primarily in the coal sector in addition to small reductions in natural gas, nuclear, and oil power plants.

According to Fig. 2.36, the share of natural gas and renewable, particularly wind energy, is expected to increase significantly over the projected period, while coal, oil, and nuclear energy are expected to decrease due to retirements of some power plants and lower growth compared to other types of power generation sources.

Imports and Exports of Electricity by Canada

Canada is a net exporter of electricity. In 2014, Canada exported 59 TWh of power, a 6,2% decrease from the level reached in 2013 (63 TWh). In 2016, Canadian electricity export volumes increased by 7% and reached a

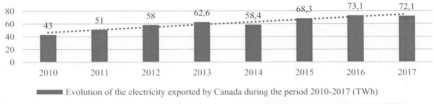

▬▬▬ Evolution of the electricity exported by Canada during the period 2010-2017 (TWh)

•••••• Linear (Evolution of the electricity exported by Canada during the period 2010-2017 (TWh))

Figure 2.37 Evolution of the electricity exported by Canada during the period 2010–17. *(Source: National Energy Board Commodity Statistics (2018).)*

new high of 73,1 TWh with respect to 2015 (68 TWh). In 2017, the electricity exported by Canada reached 72,1 TWh, a decrease of 1,4% with respect to the level reported in 2016.

The evolution of the electricity exported by Canada during the period 2010–17 is shown in Fig. 2.37.

Considering the data included in Fig. 2.37, the following can be stated: electricity exported from Canada during the period 2010–17 increased by 67,7% rising from 43 TWh in 2010 to 72,1 TWh in 2017. Except the years 2014 and 2017, all the other years within the period considered have registered an increase in the level of exports of electricity from Canada with respect to the previous year. Based on the publicly available information, it is expected that the electricity exports from Canada will continue to grow over the next years. Most of the electricity that Canada will export over the coming years will go, as in the past, to the US.

Within Canada, Quebec remained the country largest electricity exporter province, followed by Ontario, British Columbia, and Manitoba.[21] In 2016, these four provinces accounted for 95% of total Canadian electricity exports, all of which go to the US. The primary markets for Canadian electricity exports in the US are the following: New York, California, Vermont, Minnesota, North Dakota, Michigan, and Maine.

Regarding the revenue from electricity exports, it dropped 6% in 2016 after reaching a seven year high in 2015. The main reason for this decrease was the lower US wholesale prices in destination markets. In 2016, Canada earned around US$40 per MWh for its electricity exports; this represents a reduction of US$6 per MWh registered in 2015 (US$46 per MWh).

---

[21] It is important to highlight the fact that the Canadian provinces that export large amounts of electricity are usually those with predominantly hydro-based generation.

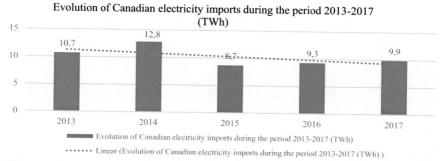

Figure 2.38 Evolution of Canadian electricity imports during the period 2013–17. *(Source: Electricity Trade Summary, National Energy Board (2018).)*

On the other hand, it is important to note that Canada consumes almost all of the electricity generated within the country to satisfy its energy needs and, for this reason, the electricity exports is less than 10% of total power generated in the country. At the same time, Canada also imports a small amount of electricity from the US.

The evolution of Canadian electricity imports during the period 2013–17 is shown in Fig. 2.38.

Considering the data included in Fig. 2.38, it can be stated that the electricity imports by Canada during the period 2013–17 decreased by 7,5% falling from 10,7 TWh in 2013 to 9,9 TWh in 2017. It is important to stress that during the period 2015–17, the electricity imports increased by 13,8% and this trend is expected to continue without change at least during the coming years. In 2015, the electricity imports were the lowest volumes in the last 20 years. The peak in electricity imports during the period considered was reached in 2014.

In general, electricity imports by Canada help avoids the costs of building additional power generation plants in the country. Also, "Canadian provinces have a greater capacity to exchange electricity with American states along north-south interconnections than between neighboring Canadian provinces. Therefore, although electricity pricing in US markets is usually higher than in Canadian markets, provinces frequently imports electricity from the US when domestic supply is limited, transmission capacity from other Canadian jurisdictions is constrained, or at times of the day when prices in the US are low" (Electricity Trade Summary, 2018).

Summing up the following can be stated: in 2040, the Canadian energy mix is expected to be as indicated in Fig. 2.39.

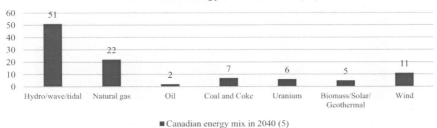

Figure 2.39 Canadian energy mix in 2040. *(Source: About Electricity, National Resources of Canada (2016).)*

According to Fig. 2.39, it is expected that conventional energy sources will increase its participation in the energy mix of the country in 2040 compared with the composition of the energy mix in 2014: natural gas will increase by 7% rising from 15% in 2014 to 22% in 2040. Coal and crude oil will reduce their role in Canada's energy mix in 2040 compared with 2014. The participation of coal in the energy mix of the country will be only 2%, and crude oil 7%.

The Future in the Use of Crude Oil for Electricity Generation in Canada

Analyzing the data included in Canada's Energy Future (2013): Energy Supply and Demand Projections to 2035 report, the following can be stated: sufficient energy supplies will be available to meet Canada's growing energy needs until at least 2035. Over the next two decades, the National Energy Board projects energy production levels that are higher than domestic needs, so it is anticipated an increase in the quantities of energy that will be available for export during the coming years.

The abundant hydroelectric resources within the country account for most of Canada's energy matrix. Considering the reserves of crude oil and natural gas, as well as the projected prices of raw materials, among other elements, the National Energy Board anticipates that the total Canadian energy production will grow substantially in the future. Crude oil production leads this growth and is expected to reach 5,12 million barrels per day in 2030, an increase of 4,9% with respect to 2025, and 6,3 million barrels per day in 2040, an increase of 28,6% with respect to 2015. The production of oil sands constitutes the major part of this foreseeable increase.

On the other hand, it is important to highlight that the Canadian electricity supply is also expected to grow steadily during the period analyzed. The total electricity generation in Canada is expected to increase by 27% up to 2040.

"Canada's energy resources are among the largest in the world, and this situation will continue to be so during the coming decades. Also, Canada ranks third globally in proven crude oil reserves, almost 97% of which are in the oil sands, and 15th in both proven natural gas and coal reserves" (BP Statistical Review of World Energy, 2015, 2016). "Canada also ranks fourth in identified resources of uranium" (Uranium, 2014 – Resources, Production and Demand, 2015), and has the same rank among global producers of crude oil (4,8% of the total). This large and diversified resource base contributes to Canada's status as an important world energy producer and exporter. Regarding production, "Canada ranks among the top five in the world for hydroelectricity, crude oil, natural gas and uranium" (BP Statistical Review of World Energy, 2015, 2016).

The Canadian energy system is in a constant state of change. Several factors such as technology, macroeconomics, infrastructure, and government policy and programs, among others, continually influence how energy is produced, transported, and consumed in Canada (Canada's Energy Future, 2016, 2016). Due to this dynamic, the following can be stated:

1. The evolution of crude oil prices is one of the many uncertainties affecting Canada's long-term energy prospects.
2. Energy production is projected to grows faster than its consumption, so net energy exports are expected to increase in the future. Crude oil production is expected to reach more than six million barrels per day by 2040.
3. The levels of future crude oil and natural gas production will depend in no small extent on the future prices of both types of energy.
4. Without the construction of new crude oil pipelines, it is probable that crude oil production in Canada will continue growing but at a moderate rate during the coming years.
5. The total energy use in Canada, which includes the use of energy in the energy production sector, is expected to grow at similar rates in all possible scenarios that have been analyzed by the competent authorities of the country.

Crucial several industry challenges are tempering long-term growth prospects in the use of crude oil for electricity generation in Canada. Some of these challenges are, among others, the following:

1. **Uncertainty**: Canada's policies and regulations in the oil sector are becoming increasingly more stringent and costlier, resulting in reduced attractiveness for investment for the development of this sector within the Canadian economy.

2. **Cumulative impacts of government policy changes**: Developing resources responsibly to help achieve critical regulatory, social, and environmental outcomes, is essential and needs to be done by the Canadian government in a manner that does not unnecessarily burden the crude oil industry and risk more jobs.

3. **Potential divergent policies from US competitors**: US oil producers may not have to face similar guidelines to those in Canada. Additionally, protectionist policies that may be pursued by the current US administration are also a cause for concern for the Canadian energy authorities, considering the level of energy trade between the two countries.

CHAPTER 3

Current Status and Perspective in the Use of Natural Gas for Electricity Generation in the North America Region

Contents

Conventional Energy in North America
ISBN 978-0-12-814889-1
https://doi.org/10.1016/B978-0-12-814889-1.00003-6

General Overview

Natural gas[1] is the second largest energy source in world power generation and the only fossil fuel whose share of primary energy consumption is projected to grow during the coming years. This type of energy source is the number three fossil fuel, representing 24% of the global primary energy, and the second energy source in power generation, representing 22% of the world total (World Energy Resources 2016, used with the permission of the World Energy Council, www.worldenergy.org, 2016). The natural gas market in the North America region is a mature market, with a financial structure well integrated and with a high national production in Canada and the US. In this type of market, the price of natural gas has been driven by the level of its supply and demand.

It is important to single out that natural gas markets operate in three regional markets—North America, Europe, and Asia—within which 90% of natural gas trade is carried out. However, these three regional markets are in a transition period from a clear regional separation market to a global market with two clear regions. These regions are the Atlantic and the Asia Pacific.

A report from MTI published in 2010, entitled "The Future of Natural Gas," estimated the remaining recoverable natural gas reserves at world level to be 16.200 trillion cubic feet, 150 times the current annual global gas consumption, with low and high projections of 12.400 trillion cubic feet and 20.800 trillion cubic feet, respectively. These natural gas reserves increase the importance of this type of energy source in the energy mix of many countries.[2]

For the above reason, natural gas continues to be an attractive energy source for the electric power and industrial sectors in many countries, including Canada and the US, because of low capital costs, favorable heat rates, and relatively low energy cost in comparison with other types of energy sources, particularly coal and oil. The US is the country with the projected most significant increase in the production of natural gas during the period 2015—40 (20%), exceeded only by the production of natural gas in the whole Middle East.[3] However, it is important to highlight that

---

[1] The main components of natural gas are the following: methane, ethane, propane, condensate, nitrogen carbon dioxide, and hydrogen.

[2] Approximately 9.000 trillion cubic feet could be economically developed with a natural gas price at or below US$4 million British thermal units at the export point.

[3] Shale gas is responsible for the projected increase in the production of natural gas in the US during the period 2015—40. Shale gas development accounts for 50% of US natural gas production in 2015, and it is expected to increase to nearly 70% in 2040.

Figure 3.1 Natural gas power plant. *(Source: Mscalora/Wikimedia Commons.)*

despite the projected increase in the production of natural gas in the US for the period 2015—40, the output of natural gas in the US decreased by 2,3% in 2016 concerning the level reached in 2015. This fall is the first reported in the country since 2006. In Canada, natural gas manufacture in 2016 represented an increase of 1,9% with respect to 2015 and is the second increase reported since 2006. The increase in the output of natural gas in Canada is expected to come mainly from tight gas. Almost 75% of the projected increase in the total consumption between 2015 and 2040 is going to be registered in the electric power and the industrial sectors. Natural gas-intensive industries, such as chemical, refining, and primary metal industries, are expected to expand over the period 2015—40 driving industrial demand (International Energy Outlook, 2017) (Fig. 3.1).

The imports of natural gas by the North America region has historically depended on the level of the national supplies of this type of energy source, and a price based on supply and demand. The North America region is projected to become a significant exporter of natural gas by 2020 and a large consumer and producer of this specific type of energy source. However, it is essential to be aware that the market of the region is, and will continue to be, dominated by the dynamics of supply and demand in the US, which is currently the largest producer and consumer of natural gas worldwide. According to EIA sources, gross production of natural gas in the US reached 28.814.028 million cubic feet in 2017, which represents an increase of 1,2% compared to the level reached in 2016. Based on the level achieved in the production of natural gas in 2017, it is expected that the US is now ready to become a net exporter of this type of energy source before 2020.

With the increase in unconventional gas production, the growth of supply has exceeded the growth of demand since 2008, has caused an excess of regional supply, and a collapse in the price of this type of energy source. In the case of the North America region, the production of natural gas has grown due to the rise of shale gas in the US. The region produced 34.750 billion cubic feet of natural gas in 2015, which represented 27,8% of the global gas production and a growth of 3,9% from the previous year. In 2016, however, the output of natural gas in the North America region reached 31.819 billion cubic feet, a reduction of 8,5% with respect to 2015. The US was responsible for the majority of this reduction (2,3%).

It is expected that unconventional gas, and particularly shale gas, will make an essential contribution to the future supply of energy in the US as well as to the efforts of this country to reduce the environmental contamination produced by the use of coal for electricity generation and heating, in particular, the emission of carbon dioxide to the atmosphere. Studying the valuations of recoverable volumes of shale gas in the US made in recent years, the following can be confirmed: the amounts of shale gas in the US have increased significantly in the last five years. The current average projection of recoverable shale gas reserves in the US is approximately 650 trillion cubic feet.[4] Nearly 400 trillion cubic feet could be considered as economically recovered with a gas price at or below US$6 million British thermal units at the well-head (The Future of Natural Gas, 2010).

Globally, there are abundant supplies of natural gas, much of which can be extracted at a relatively low cost. Without a doubt, a low–cost, abundant natural gas and the growth of renewable energy sources are accelerating the premature retirement of baseload power plants, particularly coal and nuclear power plants in the North America region.

From Table 3.1, the following can be stated: the Middle East had the highest natural gas proved reserves in the world in 2015, followed by Europe and Eurasia. In 2016 and 2017, this situation had not changed, and the Middle East continued to be the region with the highest reserves of natural gas followed by Europe and Eurasia (see Table 3.2). North America occupied the fifth position according to its natural gas reserves. The highest natural gas production in 2015, 2016, and 2017 was registered in Europe and Eurasia, followed very closely by North America (Fig. 3.2).

---

[4] These studies show a low projection of 420 trillion cubic feet and a high projection of 870 trillion cubic feet.

Table 3.1 Natural Gas Data by Region in 2015 (billion cubic feet).

Region	Proved Reserves	Production	R/P Ratio (years)
Middle East	2.826.617,7	21.821,1	129,5
Europe and Eurasia	2.005.109,3	34.955,2	57,4
Asia and the Pacific	552.607,7	19.658,2	28,1
Africa	496.666,5	7.479,2	66,4
North America	450.326,0	34.750	13,0
Latin America and the Caribbean	268.091,0	6.302,1	42,5
Total	6.599.418,2	124.965,8	52,8

Source: World Energy Resources 2016, used with the permission of the World Energy Council, www.worldenergy.org.

Table 3.2 Natural Gas Data by Region in 2016 and 2017.

Region	Proved Reserves (trillion cubic feet) 2016	Production (billion cubic feet) 2016	Proved Reserves (trillion cubic feet) 2017	Production (billion cubic feet) 2017
Middle East	2.803,2	22.558,4	2.794,2	23.304,15
Europe and Eurasia	2.002,0	35.318,5	2.195,6	37.341,73
Asia and the Pacific	613,3	20.479,2	681,8	1.080,63
Africa	503,3	7.356,1	487,8	7.945,80
North America (including Mexico)	393,0	34.764,1	381,9	33.601,91
Latin America and the Caribbean	268,0	6.250,7	290,3	6.321,33
Total	6.582,8	126.727	6.831,6	109.595,55

Source: BP Statistical Review of World Energy 2018 (June 2018).

According to World Nuclear News (WNN, 2017), some generation technologies initially designed to operate as baseload were not intended to work flexibly, and in the specific case of nuclear energy, they do not have a regulatory regime that allows them to do so. However, advances in energy technologies have created new prospects for affordable and safe natural gas supplies. For this reason, natural gas markets are increasingly

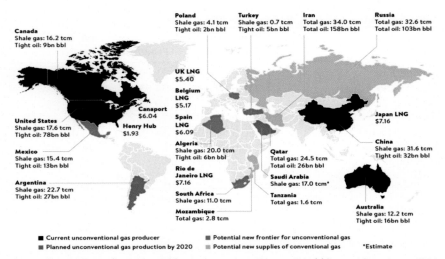

Figure 3.2 Technically recoverable gas reserves. *(Source: World Energy Resources, 2016. Used with the permission of the World Energy Council, www.worldenergy.org, 2016.)*

interconnected as a result of gas–to–gas pricing, short-term trade, and consumer bargaining power.

On the other hand, the future of energy demand is highly uncertain, and, for this reason, new policy frameworks and continuous cost improvements will be needed to make natural gas more competitive compared to other type of energy sources used for electricity generation and heating, particularly in the specific case of some renewable energy sources. Infrastructure development, government support, and the closing of regulatory gaps are necessary to unlock the socioeconomic and environmental benefits of natural gas for electricity generation and heating.

The traditional generation of baseload, mainly through the use of coal and nuclear energy, has been negatively affected by the increased use of natural gas and other renewable energies, as well as by the low growth in the electricity demand and a series of energy policy problems. "Hydropower, nuclear, coal, and natural gas power plants provide essential reliability services and fuel assurance critical to system resilience. A regular comprehensive regional and national review is needed to determine how a portfolio of domestic energy resources can be developed to ensure grid reliability and resilience" (World Nuclear News, 2017).

It is important to single out that North America represents the largest and one of the most dynamic natural gas regions in the world. Overall, North America is both the largest producing as well as consuming region of natural gas in the world. The US and Canada have large natural gas reserves, including vast reserves of unconventional gas, and these reserves will ensure the supply of gas to these two countries for many dozens of years.

"A 1,5% annual rate of growth in natural gas demand toward 2040 is healthy compared with the other fossil fuels, but markets, business models, and pricing arrangements are all in flux. A more flexible global market, linked by a doubling of trade in liquefied natural gas, supports an expanded role for natural gas in the energy global mix" (World Energy Outlook 2016, 2016). Gas consumption increases almost everywhere, with the main exception of Japan where it falls back as nuclear power is reintroduced within the country energy mix. China, where consumption grows by more than 14.126 billion cubic feet, and the Middle East are the regions with the expected most considerable growth in the use of natural gas for electricity generation and heating.

According to the World Energy Outlook 2016 report, it is expected that there will be a marked change from the previous system of stable and fixed-term relationships between suppliers and a defined group of clients, in favor of more competitive and flexible agreements, which include the greater dependence of the prices established by gas-to-gas competition. This change is motivated by the increasing availability of LNG shipments from the US, by the arrival, in the years 2020, of new gas exporters, especially from East African countries, as well as by the diversity of global supply of gas due to the continued, albeit uneven, the spread of unconventional gas revolution. The floating storage and regasification units help to identify new and smaller LNG markets, whose general participation in the long-distance gas trade is expected to grow from 42% in 2014 to 53% in 2040; this means an increase of 11% during that period.

It is important to highlight the fact that world natural gas markets are in the early stages of integration, with many obstacles to overcome for further development. If a more integrated natural gas market grows, with countries pursuing gas production and trading on an economic basis, there will be rising trade among the current regional markets, and the US could become a substantial net importer of LNG in future decades. Greater world market liquidity would be beneficial to US interests, because US prices for natural gas would be lower than under current regional markets, leading to more natural gas consumption in the country. Greater market liquidity would also contribute to security by enhancing the diversity of global supply and resilience to supply disruptions for the US and its allies. These factors are expected to moderate security concerns about import dependence.

As a result of the significant concentration of conventional gas resources at the world level, energy policy and geopolitics play, and will continue to play during the coming years, a substantial role in the development of global natural gas supply and market structures.

The Natural Gas Sector in the United States

Natural gas was the US largest source of energy production in 2016, representing between 33% and 34% of all energy produced in the country in that year. According to the EIA database (2018), in 2016, the production of natural gas in the US reached 26.662.774 million cubic feet, which is 4,7% lower than the level reported in 2015. In 2017, the production reached 26.854.288 million cubic feet, which represented an increase of 0,7% with respect to the level registered in 2016. It is important to highlight that the output of natural gas is expected to grow 6% per year until 2020, but after that year, it is projected that the production of natural gas slows down its growth to less than 1% per year until 2040. The increase in the US natural gas production during the period 2017—50 is the result of the continued development of shale and tight oil resources, which is expected to account for more than 75% of the US natural gas production in 2050 (Annual Energy Outlook, 2018).

On the other hand, in 2017, the consumption of natural gas in the US reached 27.090.166 million cubic feet, which is 1,5% lower than the level reported in 2016 (27.485.517 million cubic feet) (EIA database 2018). After 2020, natural gas production in the US is expected to grow at a higher rate than consumption in all foreseeable scenarios. Nevertheless, "to satisfy the growing demand for natural gas, production must expand into less prolific and more expensive-to-produce areas, which will put upward pressure on production costs" (Annual Energy Outlook, 2018).

Natural gas net imports (imports minus exports) by the US set a record low of 685 billion cubic feet in 2016, continuing a decline for the ninth consecutive year. Natural gas exports continued to grow; US exports in 2016 were more than three times larger than the level reached ten years ago (US Natural Gas Imports and Exports 2016, 2017). Infrastructure improvements, including natural gas pipelines and facilities for LNG for export, assisted suppliers in meeting increased demand from foreign markets. The US is expected to become a net exporter of natural gas on an average annual basis by 2018, according to the Annual Energy Outlook 2017 with projections to 2050 report.

The EIA Electric Power Annual 2016 report has stated that the net electricity generation using natural gas and other gas as fuel in the US, in 2016, reached the amount of 1.378 billion kWh, which is 3,4% higher than the level reached in 2015 (1.333 kWh). In 2017, the electricity production using natural gas as fuel reached 1.273 billion kWh, which is 7,6% lower than the level reached in 2016 and represented 31,7% of the total electricity generated in the country in that year (4.015 billion kWh). Natural gas electricity generation capacity installed in the US in 2016 was 446.823,2 MW.[5] The government and the industries have adopted measures to increase the energy efficiency, and this is one of the main reasons for the decrease in the role of natural gas in the electricity generation in the country in recent years.

The natural gas industry in the US includes exploration for production, processing, transportation, storage, and marketing of natural gas and natural gas liquids (NGL). The research for and production of natural gas and petroleum make up a single industry, and many wells in the US produce both oil and gas. It is important to know that "conventional gas is contained in porous rocks: as the gas can move around in such rocks, only a few wells need be drilled, and the gas flows for a long time. Fracking has been used since the 1950s to boost the production of conventional gas wells" (Erbach, 2014).

Natural gas that remains trapped in the rock where it was formed is called "shale gas." The production process associated with the extraction of shale gas is more complicated than the production process of natural gas. Why? The answer to this question is straightforward. According to the Erbach (2014) report, "as the gas cannot move within the rock, it is necessary to drill horizontally along the gas-containing rock, typically at a depth of 1.500—3.000 m. The rock in which the gas is trapped is then fractured by injecting a mixture of water, sand, and chemicals at high pressure. The sand serves to keep cracks open so that the gas can flow to the surface. In addition to gas, some shale gas wells produce valuable NGL (such as ethane, propane, butane, and natural gasoline), which can be very important to the economics of shale gas production".

Natural and Shale Gas Reserves in the United States

The US has a vast natural gas reserve reaching 308.500 trillion cubic feet in 2017, a 0,4% higher than the level reported in 2016. The evolution of natural gas reserves in the US during the period 2010—17 is shown in Fig. 3.3.

---

[5] The top-producing state of natural gas in the US is Texas.

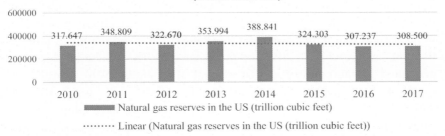

Figure 3.3 Evolution of the natural gas reserves in the US during the period 2010—17. (*Source: EIA database 2016; BP Statistical Review of World Energy 2018 (June 2018).*)

According to Fig. 3.3, this type of natural gas reserves in the US decreased by 2,9% during the whole period considered. Within the period 2010—11, this type of natural gas reserves increased 9,8%; during the period 2012—14 there was an increase of 20,5%, the highest increase within the whole period. The peak in this type of natural gas reserves in the US was reached in 2014. Between 2014 and 2015, this type of US natural gas reserves decreased by 64,5 trillion cubic feet (16,6%), declining from 388.841 trillion cubic feet in 2014 to 324.303 trillion cubic feet in 2015. During the period 2015—17, this type of US natural gas reserves decreased by 4,9%. In principle, it is expected that this type of US natural gas reserves will continue to grow moderately during the coming years, particularly in the specific case of shale gas.

In the case of dry natural gas, the evolution of the US reserves is shown in Fig. 3.4.

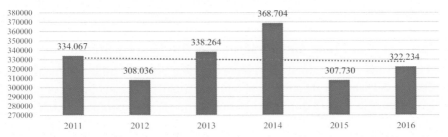

Figure 3.4 Evolution of dry natural gas reserves in the US during the period 2011—16 (trillion cubic feet). (*Source: EIA database 2018.*)

Based on the information included in Fig. 3.4, the following can be stated: the evolution of dry natural gas reserves in the US decreased by 3,6% during the period under consideration falling from 334.067 trillion cubic feet in 2011 to 322.234 trillion cubic feet in 2016. The peak in the level of the reserves of this type of natural gas during the period 2011—16 was reached in 2014.

It is important to single out also that new natural gas discoveries registered by the US in 2012 were tied to investment in onshore drilling opportunities in 48 states, specifically shale gas. Sustaining the US national gas reserves inventory is a prerequisite for continuing or growing natural gas production, which is necessary to support the future long-term market energy demand, bearing in mind that, currently, this type of energy source provides more than 90% of the natural gas consumed in the country. US natural gas reserve additions from different sources exceeded production of this type of energy source by more than 30% (US Crude Oil and Natural Gas Proved Reserves, Year-end 2016, 2018).

Summing up and taking into account the report mentioned above, the following can be stated:

- During the period 2011—12 and 2014—16, the US natural gas reserves decreased by 7,5% and 21%, respectively. In the case of dry natural gas reserves, there was a decline by 7,8% during the period 2011—12 and 17,6% during the period 2014—15.
- The average price of US natural gas in 2015 dropped by 42% compared to 2014, resulting in operators revising their 2015 natural gas proved reserves downward, just as they did with crude oil proved reserves. The US proved natural gas reserves declined by 16,6% (64,5 trillion cubic feet) in 2015 (US Crude Oil and Natural Gas Proved Reserves, 2016).
- At the state level, Pennsylvania and Oklahoma reported the most substantial net increases in natural gas proved reserves in 2016. Pennsylvania natural gas proved reserves increased by 11% (6,1 trillion cubic feet), the most significant net gain of all states in 2016. The most significant net benefits in natural gas proved reserves by volume in 2016 were reported in Oklahoma with a total of 12% (3,7 trillion cubic feet) and Ohio with 3,1 trillion cubic feet. Other increases higher than 1 trillion cubic feet also occurred in Texas, West Virginia, Louisiana, and North Dakota. The states with the most significant net decreases in natural gas proved reserves in 2016 were Alaska and New Mexico (US Crude Oil and Natural Gas Proved Reserves Year-end 2016, 2018).

Table 3.3 US Total Natural Gas Discoveries and Proved Natural Gas Reserves Wet After Lease Separation in 2016.

Source of Natural Gas	Discoveries in 2016 (trillion cubic feet)	Proved Natural Gas Reserves Wet After Lease Separation at the End of 2016 (trillion cubic feet)
Coalbed methane	—	10,6
Shale gas	32,3	209,8
Other gas	6,1	120,7
US total	38,4	341,1

Source: US Crude Oil and Natural Gas Proved Reserves, Year-end 2016, 2018.

The US total natural gas discoveries reached 38,4 trillion cubic feet in 2016 (see Table 3.3). The US proved natural gas reserves at the end of 2016 reached 341,1 trillion cubic feet.

According to Table 3.3, shale gas represents 84,1% of all discoveries of this type of natural gas in the US in 2016 and 61,2% of the total reserves reported in that year. Although the US leads the world in natural gas production, it is only fifth in proved natural gas reserves, behind Russia, Iran, Qatar, and Turkmenistan.

Based on EIA estimates the following can be stated: the share of natural gas from shale gas compared with total US natural gas proved reserves increased by 11% rising from 51% in 2014 to 62% in 2016 (see Fig. 3.5).

The top five US states with the most significant natural gas proved reserves in 2016 were Pennsylvania, Texas, West Virginia, Oklahoma, and Ohio. Pennsylvania had the highest level of natural gas proved reserves from shale gas with 60.979 trillion cubic feet, followed by Texas with 56.577 trillion cubic feet, West Virginia with 23.146 trillion cubic feet, Oklahoma with 20.327 trillion cubic feet, and Ohio with 15.472 trillion cubic feet (see Fig. 3.6).

The evolution of shale gas reserves in the US during the period 2011—16 is shown in Fig. 3.7.

Taking into consideration the information included in Fig. 3.7, it can be stated that the shale gas reserves in the US increased by 59,4% in the period 2011—16 rising from 131.616 billion cubic feet in 2011 to 209.809 billion cubic feet in 2016. It is expected that shale gas reserves in the US will continue to increase during the coming years, and will continue to be the primary sources of this increase.

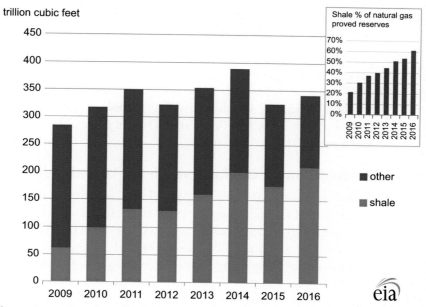

Figure 3.5 Evolution of the share of natural gas from shale gas compared with total US natural gas proved reserves during the period 2009–16.

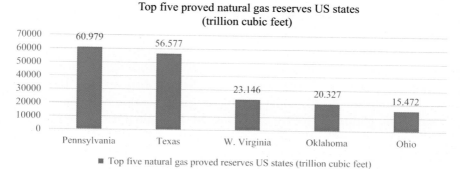

Figure 3.6 Top five natural gas reserves US states in 2016. *(Source: EIA database 2018.)*

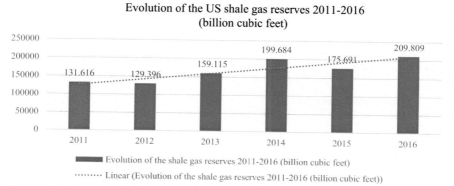

Figure 3.7 Evolution of shale gas reserves in the US during the period 2011–16. *(Source: EIA database 2018.)*

On a geographical level, Pennsylvania is, in 2016, the US state with the most natural gas proved reserves from shale gas reaching a total of 61 trillion cubic feet, followed by Texas with 56,6 trillion cubic feet, West Virginia with 23,1 trillion cubic feet, Oklahoma with 20,3 trillion cubic feet, and Ohio with 15,4 trillion cubic feet (US Crude Oil and Natural Gas Proved Reserves, Year-end 2016, 2018).

Non-associated Natural Gas Reserves in the United States

Non-associated natural gas, also called "gas well gas," is defined "as natural gas not in contact with significant quantities of crude oil in a reservoir," and includes shale natural gas and all coalbed natural gas. Proved reserves of US non-associated natural gas increased by a total of 10,1 billion cubic feet in 2016, a 4% increase in comparison with the level registered in 2015 (see Table 3.4) (US Crude Oil and Natural Gas Proved Reserves, Year-end 2016, 2018). It is predictable that the US non-associated natural gas reserves will continue to increase at least during the coming years.

Dry Natural Gas Reserves in the United States

"Dry natural gas is the volume of natural gas (primary methane) that remains after natural gas plant liquids and non–hydrocarbon impurities are removed from the natural gas stream, usually downstream at a natural gas processing plant. Not all produced gas has to be processed at a natural gas processing plant" (US Crude Oil and Natural Gas Proved Reserves, Year-end 2016, 2018). Some produced gas is dry without processing. During the period 2015—16, the estimated dry natural gas content of US total natural gas proved reserves "increased from 307,7 trillion cubic feet in 2015 to 322,2 trillion cubic feet in 2016, an increase of 5%" (US Crude Oil and Natural Gas Proved Reserves, Year-end 2016, 2018). It is projected that the US dry natural gas reserves will continue to increase at least during the coming years.

Table 3.4 Non-associated Natural Gas Proved Reserves and Changes in Reserves in the US in 2016 (billion cubic feet).

	Published Proved Reserves 2015	Adjustments	Revision Increases	Revision Decreases	Proved Reserves 2016
US	258.807	6.089	41.737	44.569	268.913

Sources: US EIA, Form EIA-23L, Annual Report of Domestic Oil and Gas Reserves 2018.

Lease Condensate and Natural Gas Plant Liquid Reserves in the United States

"Lease condensate is a mixture consisting primarily of hydrocarbons heavier than pentanes that are recovered as a liquid from natural gas in lease separation facilities. This category excludes natural gas plant liquids, such as butane and propane, which are recovered at downstream natural gas processing plants or facilities. Lease condensate usually enters the crude oil stream. Before 2015, US lease condensate proved reserves had increased for six consecutive years" (US Crude Oil and Natural Gas Proved Reserves, Year-end 2016, 2018). According to available data, US proved reserves of lease condensate declined by 16% or 472 million barrels during the period 2015—16, falling from 2.912 million barrels in 2015 to 2.440 million barrels in 2016 (US Crude Oil and Natural Gas Proved Reserves, Year-end 2016, 2018). It is projected that the US lease condensate and natural gas plant liquid reserves will continue to decrease at least during the coming years.

Natural Gas Production in the United States

In 2012, the US produced 29.542.313 million cubic feet of natural gas. In 2016, this production increased up to 32.636,50 billion cubic feet; this means an increase of 10,5% (EIA database, 2017). The total number of the US producing natural gas wells in the US in 2016 was 553.495 (EIA database, 2018).

According to EIA sources, the US has been the world's largest producer of natural gas since 2009, when US production surpassed Russia's natural gas production. In 2017, the US produced 33.177 trillion cubic feet of natural gas, an increase of 1,7% concerning the level reached in 2016.

The evolution of gross natural gas production in the US during the period 2011—17 is shown in Fig. 3.8.

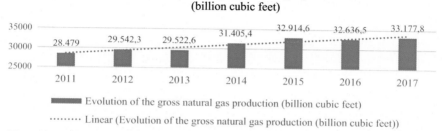

Evolution of the US gross natural gas production (billion cubic feet)

- Evolution of the gross natural gas production (billion cubic feet)
- Linear (Evolution of the gross natural gas production (billion cubic feet))

Figure 3.8 Evolution of the gross natural gas production in the US during the period 2011—17. (*Source: EIA database 2018.*)

According to Fig. 3.8, the gross natural gas production in the US increased by 16,5% during the whole period considered. The peak in the production of gross natural gas within the period 2011—17 was reached in 2017 with an output of 33.177 billion cubic feet. Except for 2013 and 2016, the gross natural gas production within the period 2011—17 increased in each one of the remaining years. It is expected that the production of gross natural gas will continue to grow during the coming years.

Natural gas production over the period 2016—20 is projected to grow nearly at a 4% annual average. Beyond 2020, US natural gas production is projected to increase at a 1% annual average as the net export is projected to growth moderates, domestic natural gas use is expected to become more efficient, and prices are expected to rise slowly. Rising prices in the US natural gas sector is supposed to be moderated as a result of advances in oil and natural gas extraction technologies (Annual Energy Outlook 2017 with Projections to 2050, 2017). On a geographic basis, the significant increase in natural gas production took place in Louisiana with 376.625 million cubic feet, Ohio with 346.719 million cubic feet, West Virginia with 225.939 million cubic feet, and Pennsylvania with 150.525 million cubic feet.

On the other hand, an increase in the US natural gas production of 1,4% as average annual change is expected during the period 2012—40. The production is expected to grow from 24 trillion cubic feet in 2012 to 35,3 trillion cubic feet in 2040; this means an increase of 47%. The evolution in the production of natural gas in the US during the period 2012—40 is shown in Fig. 3.9.

Taking into consideration the data included in Fig. 3.9, it can be stated that the US production of natural gas during the period 2012—40 is

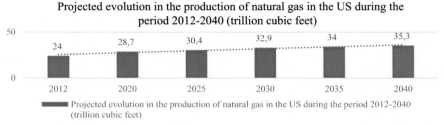

Figure 3.9 Projected evolution in the production of natural gas in the US during the period 2012—40 (trillion cubic feet). *(Source: International Energy Outlook 2016 with Projection to 2040, 2016. US Energy Information Administration, Office of Energy Analysis, US Department of Energy, IEO 2016.)*

expected to increase by 47,1% rising from 24 trillion cubic feet in 2012 to 35,3 trillion cubic feet in 2040. A significant increase is supposed to be reported during the period 2012–20 (4,7 trillion cubic feet or 19,6%). In the remaining part of the period, the maximum increase is expected to be registered between 2025 and 2030 (2,5 trillion cubic feet or 8,2%).

Finally, it is essential to highlight the following: the US production of natural gas and its participation in the US total energy production are expected to increase from 31% in 2012 to 34% in 2025, and 38% in 2040; this means that the increase in the whole period is expected to be 7% (Annual Energy Outlook, 2014). The production of shale gas and associated gas from tight oil plays is the most significant contributor to the US natural gas production growth in the period considered, expected to account for nearly two-thirds of the total US production by 2040. In 2016, the output of shale gas reached 17 trillion cubic feet, an increase of 11,8% with respect to the level reached in 2015 (15,2 trillion cubic feet) (US Crude Oil and Natural Gas Proved Reserves Year-end 2016, 2018).

On the other hand, tight gas production is the second-largest source of domestic natural gas supply in the US, but its share is projected to fall through the 2020s as the result of growing development of shale gas and tight oil plays in the country (Annual Energy Outlook 2017 with Projections to 2050, 2017).

Non-associated Natural Gas Production in the United States

Estimated production of US non-associated natural gas during the period 2014–15 increased by 1%, rising from 22,8 trillion cubic feet in 2014 to 23,1 trillion cubic feet in 2015. The most substantial increase in non-associated natural gas production was registered in Pennsylvania (Marcellus Shale), where annual non-associated natural gas production increased from 4,2 trillion cubic feet in 2014 to 4,8 trillion cubic feet in 2015, an increase of 14,3% (US Crude Oil and Natural Gas Proved Reserves, Year-end 2016, 2018). During the period 2015–16, the US production of non-associated natural gas decreased by 2%, falling from 23,1 trillion cubic feet in 2015 to 22,7 trillion cubic feet in 2016. In that year, the most substantial increases in non-associated natural gas production were registered in Ohio (Utica Shale) and Pennsylvania (Marcellus Shale). The most significant decrease in 2016 non-associated natural gas production (0,6 trillion cubic feet) was recorded in Texas (US Crude Oil and Natural Gas Proved Reserves Year-end 2016, 2018).

Dry Natural Gas Production in the United States

During the period 2011−17, dry natural gas[6] production in the US increased from 22.901.879 million cubic feet in 2011 to 26.854.288 million cubic feet in 2017 or 17,3% (see Fig. 3.10).

The evolution in the production of dry natural gas in the US during the period 2011−17 is shown in Fig. 3.10.

The peak in the production of dry natural gas during the period considered was reached in 2015. During the period examined, the output of dry natural gas decreased only in 2016 with respect to the previous year. Based on the available information, US dry natural gas production is expected to increase during the period 2017−40 from 26,8 trillion cubic feet in 2017 to 42,1 trillion cubic feet in 2040, while average annual US natural gas prices (in 2015 US dollars) remain at about US$5 per million British thermal units. Although natural gas prices remain relatively low and stable, projected development of natural gas resources in shale gas[7] and tight oil plays, tight gas, and offshore is expected to increase as a result of abundant domestic resources and technological improvements (Annual Energy Outlook, 2016).

Offshore Natural Gas Production in the United States

After decreasing during the period 2015−16 to around 1,52 trillion cubic feet, the US offshore natural gas production is expected to remain stable until 2020. After that year, it is likely to fall to 1,2 trillion cubic feet in 2023 (21% decrease), reflecting declines in production from legacy offshore fields (Annual Energy Outlook, 2018). After 2027, as increased production from new discoveries is expected to offset the decline in legacy fields, offshore

---

[6] Dry natural gas is almost completely methane. The higher the methane concentration within the gas, the drier it is. According to the EIA, dry natural gas is what remains after all of the liquefied hydrocarbons (hexane, octane, etc.) and nonhydrocarbon (helium, nitrogen, etc.) impurities are removed from the natural gas stream. On the other hand, wet natural gas contains less than 85% methane and has a higher percentage of liquid natural gasses (LNG's) such as ethane and butane. The combination of LNGs and liquefied hydrocarbons gives it the "wetness."

[7] US shale gas production is expected to grow from 10 trillion cubic feet in 2012 to 20 trillion cubic feet in 2040, an increase of 100%, offsetting declines in the production of natural gas from other sources. In 2040, shale gas is expected to account for 55% of total US natural gas production, tight gas is expected to account for 20%, and offshore production from the lower 48 states is expected to account for 8%. The remaining 17% is expected to come from coalbed methane, Alaska, and other associated and nonassociated onshore resources in the lower 48 states (International Energy Outlook, 2017).

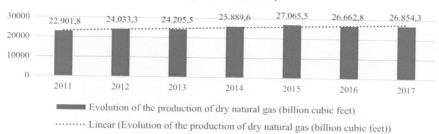

Figure 3.10 Evolution of the production of dry natural gas in the US during the period 2011–17. *(Source: EIA database 2018.)*

natural gas production is projected to increase to 1,7 trillion cubic feet in 2040; this means an increase of 41,7%. If the whole period 2016–40 is considered, then the offshore natural gas production is expected to increase by 21,4%.

Natural Gas Production and Its Technological Improvements in the United States

Natural gas possesses remarkable qualities. Among the fossil fuels, it has the lowest carbon intensity, emitting less carbon dioxide per unit of energy generated than other fossil fuels such as oil and coal. It burns cleanly and efficiently, with very few non-carbon emissions. Unlike oil, the use of natural gas as fuel in a power plant generally requires limited time to allow this type of power plant to be ready for its end-use. These favorable characteristics have enabled natural gas to penetrate many markets, including domestic and commercial heating, multiple industrial processes, and electrical power markets.

Natural gas also has desirable characteristics concerning its development and production. The high compressibility and low viscosity of natural gas allow high recoveries from natural gas conventional reservoirs at a relatively low cost. It is important to highlight that this type of fossil fuel has the characteristic to be also economically recovered from even the most hostile subsurface environments, as recent developments in shale gas formations have demonstrated.

Technology improvements are expected to increase US production from tight and shale gas formations in the coming years. Growth in US natural gas resources (proved reserves and technically recoverable resources)

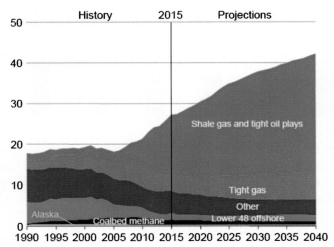

Figure 3.11 US shale gas and tight gas production during the period 1990−2040 (trillion cubic feet). *(Source: Annual Energy Outlook 2016, 2016. US Energy Information Administration.)*

and cumulative production have averaged 2,5% per year for natural gas during the period 1990−2005, and 3,1% per year during the period 2005−15. These technology improvements have allowed, and are likely to continue to permit, the expansion of tight and shale gas production in the US (Annual Energy Outlook, 2016). "However, there are risks associated with the production of shale oil and shale gas. In particular, several environmental risks are drawing lots of attention from both environmentalists and the industry. Many in the industry, along with environmentalists, are very worried about water contamination. There is the risk that, in the process of drilling so many wells, the water could be contaminated. There are also risks that increased drilling could lead to increased quantities of methane leaking from the drilling, which could add to greenhouse gas emissions" (Nyquist, 2018).

According to Fig. 3.11, the significant increase in the production of tight and shale gas in the US during the period 2015−40 will come from shale gas, including also tight oil plays. It is expected that this situation will not change significantly after 2040.

Natural Gas Consumption in the United States

The US, according to the International Energy Outlook 2016 report, "is the world's largest consumer of natural gas, and leads the North America

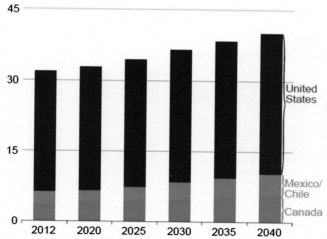

Figure 3.12 Natural gas consumption in Canada and the US during the period 2012–40. *(Source: International Energy Outlook 2016, May 2016. Energy Information Administration (EIA).)*

region in annual natural gas consumption growth with an increase of 4,2 trillion cubic feet from 2012 to 2040"; this represents 51% of the region's total increase in natural gas consumption (see Fig. 3.12).

According to the Annual Energy Outlook 2016 report, the US natural gas consumption is expected to grow from 27,5 trillion cubic feet in 2015 to 34,4, trillion cubic feet in 2040; this means an increase of 25% for the whole period under consideration. The consumption of natural gas for electricity generation and heating in the US up to 2040 is expected to increase by 2,4 trillion cubic feet, accounting for 34% of the total increase expected to be registered during the period considered. The peak in the consumption of natural gas in the US was reached in 2015, due to low natural gas prices and the retirement of coal-fired power plants capacity.[8]

According to Fig. 3.12, the US is by far the country with the highest natural gas consumption within the North America region, and the country

---

[8] The consumption of natural gas in the US for electricity generation and heating is expected to decline during the period 2015–2021 as a result of the foreseeable rising of natural gas prices and increasing use of renewable energy sources. During the period 2012–30, the consumption of natural gas in the US is expected to grow by an average of around 4% per year, and under 1% per year during the period 2031–40. This situation is the result of the reduction or phasing out of some renewable tax credits and relatively low natural gas prices (Annual Energy Outlook 2016, 2016).

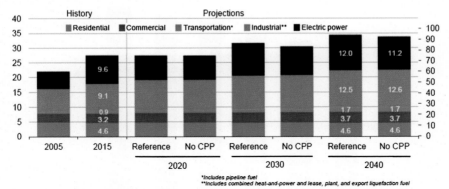

Figure 3.13 Dry natural gas consumption growth in the US during the period 2005–40. *(Source: EIA Annual Energy Outlook 2016. Presentation by Adam Sieminski, Administrator, Independent Statistics & Analysis, www.eia.gov.)*

to register the most significant increase in the use of dry natural gas during the period 2020–40 (Fig. 3.13).

According to Fig. 3.13, the following can be stated: in 2015, the majority of the use of dry natural gas in the US was for electricity generation and heating with 9,6 trillion cubic feet, followed very closely by the industrial sector with 9,1 trillion cubic feet. The consumption of dry natural gas in the industrial sector, which included the use of natural gas for lease and plant fuel and LNG for export, is expected to increase by 3,4 trillion cubic feet during the period 2015–40, an average increase of 1,3% per year.[9] It is projected that in 2040, the industrial sector with 12,5 trillion cubic feet will overcome the use of dry natural gas for electricity generation and heating (12 trillion cubic feet). However, no change is predictable in the structure of the consumption of dry natural gas in the residential, commercial, and transportation sectors in the US during the same period. On the other hand, the use of natural gas in the transportation sector currently accounts for only a small portion of US total consumption of dry natural gas, but it is expected to grow rapidly from 64 billion cubic feet in 2015 to 658 billion cubic feet in

[9] "Energy-intensive industries and those that use natural gas as a feedstock, such as bulk chemicals, benefit from relatively low natural gas prices throughout the coming years. Increasing use of lease and plant fuel, which is correlated with natural gas production, and fuel used for the production of LNG for export also contribute to the growth of natural gas consumption in the industrial sector" (Annual Energy Outlook 2016, 2016).

2040; this means an increase of more than 10 times. "Heavy-duty vehicles and freight rail account for 33% of the natural gas used in the transportation sector in 2040, and pipeline compressor stations account for most of the remainder" (Annual Energy Outlook 2016, 2016).

In the US, it is anticipated that the consumption of natural gas in the industrial sector will grow by 1,4 trillion cubic feet during the period 2012−20, with 1,3 trillion cubic feet (97%) added in the country, where industrial consumption is expected to increase by an average of 1,8% per year. The growth in the use of natural gas in the US industrial sector is likely to be low during the period 2020−40 averaging 0,5% per year and increasing by a total of 1 trillion cubic feet over that period (International Energy Outlook 2016, 2016).

In the residential sector, the consumption of natural gas for space heating is expected to decline, partially as a result of improvements in energy efficiency and migration of population to warmer regions. In the commercial sector, where growth in floor space more than offsets improvements in energy efficiency, the consumption of natural gas is expected to rise gradually over the projection period (Annual Energy Outlook 2016, 2016).

It is important to highlight that with the consideration of the proposed CPP, US natural gas consumption would be 1,7 trillion cubic feet higher in 2020 compared to the level of the consumption included in the International Energy Outlook 2016 report used as a reference. Most of the increase in natural gas consumption would occur in the electric power sector as a result of the closure of old and ineffective coal-fired power plants and the transformation of this type of plants to natural gas-fired power plants. After 2020, the effect of the CPP on natural gas use in the power sector is expected to decrease at the same time as generation from renewable energy increases. In 2040, projected US natural gas consumption is 1 trillion cubic feet lower with the CPP than in the report mentioned above. Effects of the final CPP on natural-gas-fired power generation will depend on natural gas prices, renewable technology costs, and state level implementation decisions. "An increase in natural gas use through 2040 is certainly possible in scenarios with low gas prices and implementation strategies that favor gas in comparison to other energy sources" (International Energy Outlook 2016, 2016).

The evolution of the consumption of natural gas in the US during the period 2011−17 is shown in Fig. 3.14.

According to Fig. 3.14, the consumption of natural gas in the US during the period 2011−17 increased by 10,7% rising from 24.477,40 billion cubic

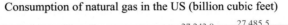

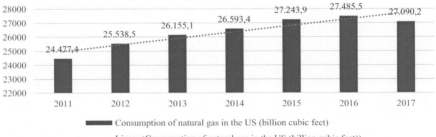

Figure 3.14 Evolution of the consumption of natural gas in the US during the period 2011–17. *(Source: EIA database 2018.)*

feet in 2011 to 27.090,20 billion cubic feet in 2017. In each year within that period, the consumption of natural gas increased concerning the previous year, except for the period 2016–17 where the use of natural gas decreased by 1,5%. The peak in the consumption of natural gas within the period considered was registered in 2016. It is expected that the use of natural gas in the US will continue to increase but moderately during the coming years.

The evolution of the consumption of natural gas for electricity generation and heating in the US during the period 2010–17 is included in Table 3.5.

Based on the data included in Table 3.5, the following can be stated: the use of natural gas for electricity generation and heating in the US increased by 22,9% during the period considered rising from 7.680.185 million cubic feet in 2010 to 9.440.777 million cubic feet in 2017. It is predictable that the use

Table 3.5 Evolution of the Consumption of Natural Gas for Electricity Generation and Heating in the US During the Period 2010–17.

Year	Total Natural Gas Consumed (million cubic feet)
2010	7.680.185
2011	7.883.865
2012	9.484.710
2013	8.596.299
2014	8.544.387
2015	10.016.576
2016	10.170.110
2017	9.440.777

Source: EIA database 2018.

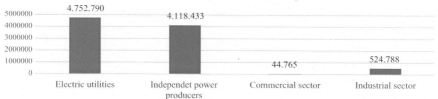

Figure 3.15 The consumption of natural gas in the US electric power sector in 2017. *(Source: EIA database 2018.)*

of natural gas for electricity generation and heating in the US will continue to increase during the coming years, as a result of the decision of the government and the power industry sector to close more old and ineffective coal-fired and nuclear power plants with many years of operation.

The consumption of natural gas in the US electric power sector in 2017 is shown in Fig. 3.15.

According to Fig. 3.15, the consumption of natural for electricity generation and heating in 2017 in the electric utility sector was the highest concerning all other components of the US electric power sector with 50,3% of the total electricity generated in the country using this type of energy source as fuel. The second sector with a high percentage in the consumption of natural gas for electricity generation and heating was the independent power producers with 43,6%.

Considering the new energy policy adopted in 2017 by the Trump administration, the use of natural gas for electricity generation and heating despite decreasing in the near-term is predictable to increase during the coming years, raising its participation in the energy mix of the country.

Finally, it is important to note that natural gas prices are likely to grow discreetly during the period 2020—2030 as electric power consumption is probable to increases; however, natural gas prices are expected to stay relatively flat after 2030 as technology improvements keep pace with expected rising demand.

Exports of Natural Gas and Liquid Natural Gas from the United States

The volume of natural gas trade flow and LNG exports from the US depends on the difference between the natural gas price in the US and the natural gas price in the rest of the world. The size of natural gas reserves in the US and the pace of technological improvements introduced to raise the efficiency of the sector, undoubtedly affecting the ability of US producers to supply the natural gas required to meet national needs, and produce an increase in the cost of domestic supplies of this type of energy source. Low world oil prices reduce the competitiveness of LNG in world markets, while exports to Canada and Mexico are directly affected by US natural gas prices. When the prices of natural gas in the US rise, exports of this type of energy resource decrease, and when the price falls, natural gas exports increase[10].

With higher international natural gas prices, particularly in Asia, US LNG exports are more competitive. The higher growth in US LNG exports increases the call on domestic production, which in turn leads to higher local natural gas prices. The increased demand for US LNG exports is offset somewhat by lower natural gas exports to Canada and Mexico as prices rise. According to the Annual Energy Outlook 2016 report, US exports of natural gas is expected to increase to 8,9 trillion cubic feet to 2040, and US LNG exports are expected to rise to 6,7 trillion cubic feet. In another scenario where there is less incentive for US LNG, total US exports of natural gas are expected to grow only to 6,8 trillion cubic feet in 2040, with US LNG exports of 5,6 trillion cubic feet.

It is well-known that lower production costs lead to more natural gas production. With assumptions of a larger resource base and more rapid improvement in production technologies, the US is expected to become a net exporter of natural gas before 2020 with net exports averaging 0,6 billion cubic feet per day (Short-Term Energy Outlook, 2017). US LNG exports are expected to increase to 10,3 trillion cubic feet in the period 2035−40. In another scenario, US natural gas production is probably to be smaller because of a smaller resource base and a slower improvement in technology. In this case, US natural gas exports are supposed to be 4,7 trillion cubic feet in 2020, with US LNG exports expected to reach 2,3 trillion cubic feet in that year, and remain at roughly the same level through 2034 before declining slightly through 2040 (Annual Energy Outlook 2016, 2016) (Fig. 3.16).

---

[10] Natural gas is the fossil fuel with the highest global demand due to its positive impact on the reduction of environment pollution among the three main fossil fuels. Undoubtedly, more natural gas exports from the US will bring environmental and strategic political relief to other countries, while simultaneously providing to the US certain economic and political benefits (Taylor, 2016).

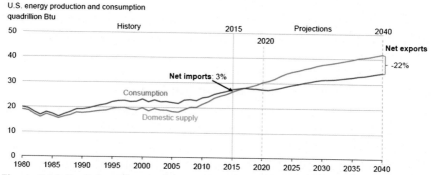

Figure 3.16 Evolution of natural gas production and consumption in the US during the period 1980—2040. *(Source: EIA Annual Energy Outlook 2016. Presentation by Adam Sieminski, Administrator, Independent Statistics & Analysis, www.eia.gov.)*

Finally, it is important to highlight that natural gas exports are expected to continue to grow in the US during the coming years. US exports in 2016 were more than three times larger than the level reached 10 years ago. Infrastructure improvements, including natural gas pipelines and facilities for LNG for exports, assisted suppliers in meeting increased demand from foreign markets. Without a doubt, the most substantial volumes of natural gas traded internationally by pipelines currently are between Canada and the US. For this reason, the US is expected to become a net exporter of natural gas on an average annual basis before 2020, according to the Annual Energy Outlook (2017) report. However, it is important to highlight that the natural gas exports through pipelines from Canada to the US are expected to decline as a result of the increase in the production of shale gas in the US (International Energy Outlook 2016, 2016).

Based on the data included in Table 3.6, the following can be stated: the US natural gas exports increased by 44,9% during the period under

Table 3.6 Evolution of the US Natural gas Exports During the Period 2011—16 (billion cubic feet).

	Five-years Average (2011—15) in the exports of Natural Gas	2015	2016	2016 Versus 2015 (%)
Pipeline	1.569,5	1.754,9	2.128,3	21
LNG	29,2	28,4	186,8	558
LNG reexports	4,3	11,6	2,6	−78
CNG	0,1	0,2	0,2	−3
Total	1.603,1	1.795,1	2.317,9	30

Source: Natural Gas Imports and Exports, 2016. US Energy Information Administration (EIA).

consideration rising from 1.598,9 billion cubic feet during the period 2011—15 to 2.315 billion cubic feet in 2016. The major growth was reported by the exports of LNG, which increased 558% during the period under consideration.

Imports of Natural Gas by the United States

According to the Natural Gas Imports and Exports 2016 report, natural gas imports by the US, in 2016, increased by 10% rising to 3.006 billion cubic feet. More than 97% of US natural gas imports come through pipelines from Canada. In the case of Mexico, natural gas imports by the US were less than 1 billion cubic feet in 2016.

On the other hand, LNG imports by the US decreased by 3% falling to 88 billion cubic feet in 2016, the second-lowest level registered since 1998. About 95% of the US LNG imports are from Trinidad. The US also imported a small amount of compressed natural gas (CNG) from Canada in 2016, about the same level as in 2014 and 2015. Natural gas net imports set a record low of 685 billion cubic feet in 2016, continuing a decline for the ninth consecutive year (Natural Gas Imports and Exports 2016, 2016).

The evolution of the imports of natural gas by the US during the period 2011—17 is shown in Fig. 3.17.

Considering the data included in Fig. 3.17, it can be stated that the imports of natural gas by the US during the period 2011—17 decreased by 12,3% falling from 3.468,70 billion cubic feet in 2011 to 3.042,40 billion cubic feet in 2017. The lowest level of imports of natural gas by the US during the period considered was registered in 2014. After that year, the imports of natural gas by the US increased by 12,9%.

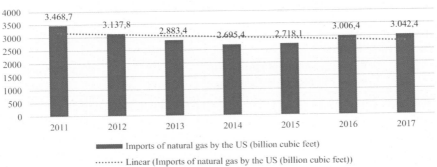

Imports of natural gas by the US (billion cubic feet)

Figure 3.17 Evolution of the imports of natural gas by the US during the period 2011—17. (*Source: EIA database 2018.*)

In recent years both US natural gas production and consumption have improved, although production has grown slightly faster, reducing the reliance on natural gas imports and reducing domestic prices. However, it is projected that the US will continue importing natural gas at least during the coming years but at a lower volume, to satisfy its domestic needs at a better price.

According to the EIA database (2018), the major natural gas supplier to the US in 2017 was Canada with 97,4% (2.962,7 billion cubic feet).[11] The US imported 78.011 million cubic feet in 2017 from Trinidad (70.450 million cubic feet), Nigeria (5.992 million cubic feet), and Canada (1.569 million cubic feet). The US also imported, in 2017, a total of 345 million cubic feet of CNG from Canada.

Unconventional Gas in the United States

Unconventional gas is the collective term used for shale gas, tight gas, and coalbed methane.[12] It is important to single out that the US and Canada are the only two countries that produce natural gas and oil from shale formations on a commercial scale; other countries have mainly exploratory test wells, except China that has recently begun commercial production from shale formations located within its territory.

Technological innovation and productivity gains have unlocked vast resources of tight oil and shale gas in the US. As a result of this situation, US shale gas is expected to grow by around 4% per annum during the period 2014—35, to account for around three-quarters of the total US natural gas production in 2035 and reach approximately 20% of the global output. "The past surprises in the strength of the shale revolution underline the considerable uncertainty concerning its future growth" (BP Energy Outlook Focus in North America, 2016).

The ongoing controversy in the US over the most effective method of extracting unconventional oil and gas from hard-to-reach underground reservoirs could curb the anticipated use of these types of energy sources in the future in the country. In a recent analysis conducted by a research team led by the Cornell University and published in the Proceedings of the National Academy of Sciences, the team examined reports of more than

---

[11] It is expected that the volume of imports of natural gas by the US from Canada will be lower than the level reached during the 2010s, as a result of the increase in the production of this type of energy source by the US, at least during the coming years.

[12] Both shale gas and conventional gas are natural gas, consisting mainly of methane.

41.000 conventional and unconventional oil and natural gas wells in Pennsylvania drilled between January 2000 and December 2012. The report provoked a strong reaction in favor of and against its content (Ingraffea et al., 2014).

Natural Gas Pipelines in the United States

The US natural gas pipeline system is a complex structure that is used to carry natural gas nationwide. "The US has more than 217.000 miles of interstate natural gas pipelines to deliver natural gas from producing regions to end users" (Estimated Natural Gas Pipelines Mileage in the Lower 48 States, 2008). The system is also used to import and export this type of energy source for its use by millions of people daily to satisfy their consumer and commercial needs (Natural Gas Pipeline System in the United States, 2018). "Across the country, the US natural gas pipeline network has about three million miles of mainline and other pipelines that link natural gas production areas and storage facilities with consumers. This natural gas transportation network delivered more than 25 trillion cubic feet of natural gas in 2016 to about 74 million customers" (Natural Gas Pipelines, 2017) (Fig. 3.18).[13]

Of the lower 48 US states, those with the most natural gas pipelines running through them are, according to the EIA source, Texas (58.588

---

[13] According to Natural Gas Pipelines (2017), approximately half of the main gas pipeline network in the US and a large part of the local natural gas distribution network was installed during the period 1950–60, when the demand for natural gas consumption increased more than doubled after the end of World War II. In the 1990s, low natural gas prices stimulated the rapid construction of natural gas-fired power plants and, for this reason, about 225.000 miles of new gas pipelines with local distribution were built to service new commercial and housing facilities constructed during that decade. Approximately 34.260 more miles of gas pipelines were built during the period 2000–14. In 2003, natural gas passed coal as the energy source with the largest installed electricity generation capacity in the US.

Natural gas prices increased substantially up to 2008. Higher prices encouraged, in turn, natural gas producers to expand existing fields and to begin exploration of new previously undeveloped natural gas fields. On the other hand, advances in drilling and production techniques led to increases in the production of shales and other geological formations that were previously not possible to exploit. These increases in the production of natural gas contributed to the general decrease in its prices since 2009, which in turn contributed to the increase in its demand for electricity generation and heating and for its use in the industry. Due to the decrease in price and the increase in the production of natural gas, it was necessary to build new gas pipelines to link the new sources of natural gas production with new consumers throughout the country (Natural Gas Pipelines, 2017).

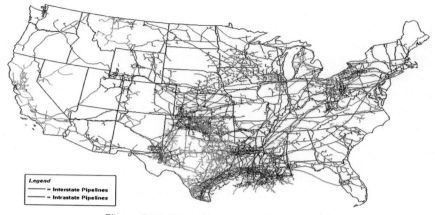

Figure 3.18 Natural gas pipelines in the US.

miles), Louisiana (18.900 miles), Oklahoma (18.539 miles), Kansas (15.386 miles), Illinois (11.900 miles), and California (11.770 miles). The states with the least natural gas pipelines running through them are Vermont and New Hampshire (Natural Gas Pipeline System in the United States, 2018).

The Future of Natural Gas for Electricity Generation in the United States

Natural gas has moved to the center of the current debate on energy, security, and climate change in the US. It is anticipated that the current US energy boom will continue for decades, and natural gas will replace coal as the largest source of US electricity generation and heating by 2035, according to the US Department of Energy forecast.[14] This US energy boom is mostly due to the combined use of new drilling techniques,[15] which releases oil and gas from shale reservoirs. The natural gas boom in the US raised the prospect of lower electricity costs, and for this reason, utilities have expanded natural gas-fired capacity to seize the opportunity. For this reason, natural gas production is expected to grow steadily, jumping by 56% from 2012 to 2040, according to an early release of an annual report by the US EIA.

[14] The fastest growing use of natural gas today in the US is for electricity generation and heating. Natural gas-fired power plants are currently among the cheapest power plants to construct.

[15] These new techniques are horizontal drilling and hydraulic fracturing or fracking.

Converting coal-fired power plants to natural gas-fired power plants is a practical and sustainable opportunity, particularly given the historically low cost of natural gas in the US. Natural gas-fired power plants produce about 45% less CO_2 than do coal-fired power plants, and they produce no sulfur dioxide or mercury. But to determine whether the conversion of coal-fired power plants to natural gas-fired power plants is a viable option, the following factors, among others, should be considered:

- natural gas and coal reserves;
- natural gas pipelines capacity;
- price volatility;
- the complexity of the conversion; and
- energy efficiency: a typical coal-fired power plant has an energy efficiency of 33%, while a typical natural gas-fired power plant has an energy efficiency of 42%; in the case of a natural gas combined cycle, the energy efficiency could be as a high as 60%.

The Impact of the Increase in Natural Gas Production in the Manufacturing Sector in the United States

The impact of the economic downturn in the US has been felt in numerous areas of domestic business, but particularly in the manufacturing sector. As the domestic situation worsened in 2008 and 2009, many companies found it more financially sound to outsource manufacturing overseas, which left a significant hole in the US economy, and a reduction of the demand for energy at the national level. However, several factors have led to the recent growth of domestic manufacturing in the country. One of these factors is the increase in foreign salaries that are making more expensive to outsource than to produce the same product in the country. That trend is also being fueled by the recent US oil and gas boom, which has helped lower energy costs in almost all economic sectors. Another factor is the decision of the current US administration to impose a tax on the import of certain products manufactured by US companies in third countries.

Wholesale prices for natural gas have fallen by around 50% since 2005. Currently, natural gas costs approximately three times less in the US than it does in competitive markets such as China, France, and Germany. Prices are not expected to rise anytime soon either, as the country should continue to enjoy the economic benefits of increased natural gas production for years to come, particularly the manufacture of unconventional gas. According to experts' estimates, by the second half of the current decade, natural gas will account for anywhere between 5% and 8% of manufacturing costs for some of Japan's and Europe's biggest exporters but will represent only 2% of US manufacturing costs. In the US electricity generation sector, natural gas will account for just 1% of US manufacturing costs.

Based on what has been said in the above paragraph, it just makes more financial sense to manufacture in the US using natural gas as fuel than it did before the shale gas revolution. For this reason, and bearing in mind the new energy policy adopted by the Trump administration, companies from around the world are already adopting measures to make long-term manufacturing investments in the US and to take advantage of low-cost natural gas in the country. Previously tens of billions of dollars in new investments in the US have been announced by several large companies, and it is expected to see more such investment in the coming years.

Electricity Generation Using Natural Gas as Fuel in the United States

Undoubtedly, the US power sector is the leading growth sector using natural gas as fuel for electricity generation and heating under the current limitations of CO_2 emissions. The expansion in the use of renewable energies for electricity generation and heating, such as wind and solar power, for example, significantly affects the use of natural gas in the electricity sector. The impacts are entirely different in the short and long-term. In the short-term, the main effect of the increase in electricity generation and heating through the use of renewable energies is the displacement of electricity generation and heating from power plants with the highest variable cost, which is natural gas in most US markets, to electricity generation plants with lower variable costs. In the long-term, the increase in electricity generation and heating through the use of renewable energies will have two likely outcomes:

- greater installed capacity of flexible power plants, mostly natural gas-fired power plants, but generally with low utilization due to their higher generation cost; and
- displacement of capacity and base-load electricity generation technologies using oil, coal, or nuclear energy as fuel, to other base-load electricity generation technologies.

Projections in the Annual Energy Outlook 2016 report focus on the factors expected to shape US energy markets through 2040. According to this report, CPP requirement to reduce CO_2 emissions accelerates the shift in the energy mix and imposes additional costs on higher-emitting energy sources of this contaminating gas. Combined with lower natural gas prices and the extension of renewable tax credits, the CPP accelerates the shift toward less carbon-intensive electricity generation power plants. In the report mentioned above, which includes reference to the CPP, a total of 92 GW of coal-fired capacity is expected to be retired by 2030; a total of 32 GW more is projected to be retired by 2030 in the No CPP case. Coal-fired generation in 2040 is expected to be 32% lower than the level reached in 2015. From 2015 levels, natural gas-fired electricity generation is expected to increase by 26% in 2030 and by 44% in 2040, and production from renewables is expected to increase by 99% in 2030 and by 152% in 2040.

The CPP requirement adopted by the US Administration has the aim of reducing carbon dioxide emissions, at the same time, to accelerate the shift in the energy mix. The CPP requirement for states to develop plans to reduce CO_2 emissions imposes additional costs on higher-emitting energy sources contaminating gases, such as coal. Combined with lower natural gas prices and the extension of renewable tax credits, the CPP accelerates the shift toward less carbon-intensive electricity generation power plants.

Natural gas has been the largest source of electricity generation and heating in the US since the middle of 2015 reaching in that period the same level as coal. After 2015, the net electricity generation using natural gas is projected to increase significantly, well above the levels of net electricity generation anticipated for coal and oil up to 2040.

The low price of natural gas in the US, together with its smaller carbon footprint compared to coal, has encouraged a rapid growth in electricity produced using natural gas as fuel as can be seen in Fig. 3.19 (Energy in the United States, 2018).

The changes in the energy mix of electricity generation capacity (including central power plant and end-use generators) are affected differently by the two implementation approaches considered in the Annual Energy Outlook 2016 report. In the CPP case, 14 GW more coal-fired generation capacity is retired, and 48 GW more natural gas generation capacity is added between 2015 and 2040. According to the report mentioned above, in 2040, the situation of the electricity generation and heating in the US is expected to be the following:

- coal-fired electricity generation is expected to reach 436 billion kWh lower than in 2015;
- natural gas-fired electricity generation is likely to be 594 billion kWh higher than in 2015; and
- renewable electricity generation is expected to be 828 billion kWh higher than in 2015.

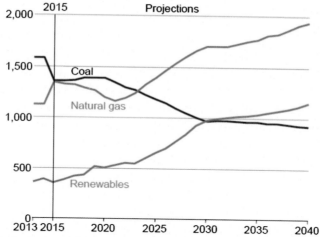

Figure 3.19 Projections of electricity generation from coal, natural gas, and renewables in the US during the period 2013–40. (*Source: Annual Energy Outlook 2016, 2016. US Energy Information Administration.*)

In the second implementation approach adopted, in 2040, the electricity generation and heating in the US is expected to be the following:

- coal-fired electricity generation is supposed to be 275 billion kWh lower than in 2015;
- natural gas-fired electricity generation is anticipated to be 375 billion kWh higher than in 2015; and
- renewable electricity generation is expected to be 898 billion kWh higher than in 2015.

Even in the absence of the CPP, the extension of renewable tax credits, as well as declining capital costs for solar photovoltaics, the adoption of other emission regulations that affect coal, and low natural gas prices, will contribute to a reduction in coal's share of total electricity generation and heating in the US. Without considering the CPP case, coal-fired electricity generation and heating is expected to change during the period 2015–40. During that period, the coal share of total electricity generation and heating is predicted to fall from 33% in 2015 to 26% in 2040; this means a reduction of 7%. Coal-fired electricity generation capacity will be restricted in the near-term by emission regulations of CO_2 and other contaminating gases to be adopted by the government. In the long-term, additions to coal-fired electricity generation capacity are expected to be limited by low natural gas prices and increased pressure from renewable electricity generation and heating.

For this reason, 60 GW of coal-fired generating capacity is supposed to be retired during the period 2016–30. On the other hand, natural gas-fired electricity generation capacity is expected to decline during the period 2016–20 in response to an increase in the use of wind and solar energy for electricity generation and heating resulting from both declining installation costs and the extension of key federal tax credits for the use of these technologies for this specific purpose. After 2020, however, the natural gas share of total electricity generation and heating is expected to increase steadily, accounting for 34% of the whole electricity generation and heating in 2040 (Annual Energy Outlook 2016, 2016).

The evolution of the consumption of natural gas for electricity generation and heating in the US during the period 2010–17 is shown in Fig. 3.20.

According to Fig. 3.20, the consumption of natural gas for electricity generation and heating in the US during the period 2010–17 increased by 22,9%. The peak in the consumption of natural gas for electricity generation and heating within the period considered was achieved in 2016. It is projected that the use of natural gas for electricity generation and heating in the US will continue to increase during the coming years, due to a probable increase in the production of this type of energy source, and despite the reduction in the use of natural gas registered in 2017 compared with 2016.

The evolution in the number of natural gas-fired power plants used in the US for electricity generation and heating is shown in Table 3.7. In December 2017, there were about 8,084 power plants in the US that have operational generators using natural gas as fuel with nameplate electricity capacities of at least 1 MW.

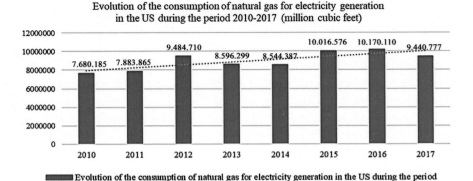

Figure 3.20 Evolution of the consumption of natural gas for electricity generation and heating in the US during the period 2010–17. (*Source: EIA database 2018.*)

Table 3.7 Evolution in the Number of Natural Gas-Fired Power Plants Used in the US for Electricity Generation and Heating During the Period 2006–16.

Year	Natural Gas-Fired Power Plants	Other Gases	Total
2006	1.659	46	1.705
2007	1.659	46	1.705
2008	1.655	43	1.698
2009	1.652	43	1.695
2010	1.657	48	1.705
2011	1.646	41	1.687
2012	1.714	44	1.758
2013	1.725	44	1.769
2014	1.749	43	1.792
2015	1.779	45	1.824
2016	1.801	45	1.846

Source: Electricity Power Annual 2016, 2018. US Energy Information Administration (EIA).

From the information included in Table 3.7, the following can be stated: the number of all gas-fired power plants used for the electricity generation and heating in the US increased by 8,3% during the period considered, rising from 1.705 power plants in 2006 to 1.846 power plants in 2016. It is anticipated that the number of gas-fired power plants in the US will continue to increase during the coming years.

According to the Electricity Power Annual 2016 (2018) report and the EIA Electricity Data Browser 2018, the net electricity generation using natural gas and other gases as fuel in 2017 in all sectors reached the amount of 1.287.053 thousand MWh, which is 7,5% lower than the level reached in 2016 (1.391.114 thousand MWh). Per sectors, the electricity generation and heating using natural gas and other gases as fuel in the US in 2017 was the following:

- Electricity utilities: the electricity generation was 617.725 thousand MWh (5,7% lower with respect to 2016) and from other gases 164 MWh (6,5% higher concerning 2016).
- Independent power producers: the electricity generation was 558.439 thousand MWh (10,6% lower with respect to 2016) and from other gases 4.013 thousand MWh (6,8% higher concerning 2016).
- Commercial sector: the electricity generation was 7.516 thousand MWh (2,9% lower with respect to 2016).
- Industrial sector: the electricity generation was 89.188 thousand MWh (2,3% lower concerning 2016) and from other gases 9.982 thousand MWh (12,2% higher concerning 2016).

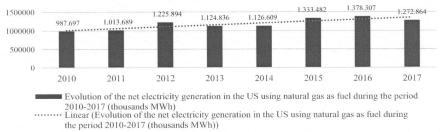

Evolution of the net electricity generation in the US using natural gas as fuel during the period 2010-2017 (thousands MWh)

■■■ Evolution of the net electricity generation in the US using natural gas as fuel during the period 2010-2017 (thousands MWh)
········ Linear (Evolution of the net electricity generation in the US using natural gas as fuel during the period 2010-2017 (thousands MWh))

Figure 3.21 Evolution of the net electricity generation and heating in the US using natural gas as fuel during the period 2010–17. *(Source: Electricity Power Annual 2016, 2018. US Energy Information Administration (EIA) and Electricity Data Browser, 2018.)*

The evolution of the net electricity generation and heating in the US using natural gas as fuel during the period 2010–17 is shown in Fig. 3.21.

According to Fig. 3.21, the net electricity generation and heating in the US using natural gas as fuel increased by 28,9% during the period considered. The peak in the electricity generation and heating during the period examined was reached in 2016. As a result of the new energy policy adopted in 2017 by the Trump administration, and taking into account the current price of natural gas, it is probable that the use of natural gas for electricity generation and heating will continue to grow during the coming years.

Summing up the following can be stated: the US has drastically increased its reserves and production of natural gas in recent years, becoming, in fact, the largest producer of natural gas worldwide. The country occupies the fifth place in the total volume of proven natural gas reserves at the world level.

However, the significant increase in natural gas production in the US is mainly due to considerable improvements in the unconventional gas extraction process. The US is, undoubtedly, a world leader in the extraction of unconventional gas, which has led to the reduction of domestic costs and an increase in the efficiency associated with the exploitation of this type of gas. For the most part, these improvements have been driven by both large and small independent producers operating in critical areas of the US where there are sizable unconventional gas reserves. Domestic consumption of natural gas has also increased as a result of an increase in domestic production and a lower price of this type of energy source. It is important to highlight that the US is the world's largest consumer of natural gas (Gas in the United States of America, 2018).

Figure 3.22 Shady hills natural gas power plant, Florida. *(Source. Courtesy Daniels Oines, Wikimedia Commons.)*

Natural gas's share within the domestic energy mix has been increasing during the last years, mainly at the expense of the reduction of coal's share in the energy mix of the country. The US has also made significant strides toward becoming a major natural gas exporter via LNG. Based on the LNG export terminals under construction, the US is projected to have five functioning terminals before the end of 2019. Once expected to be a major importer of natural gas, the combination of their increased domestic natural gas production and their LNG export pursuit has caused the EIA to project the US as a net natural gas exporter before 2020 (Gas in the United States of America, 2018).

During the period 2016–20, the US is likely to increase the number of natural gas generators used for electricity generation and heating in 281 units with an additional capacity of 54.872 MW according to EIA sources (Fig. 3.22).

The Natural Gas Sector in Canada

Canada is a country with substantial energy resources. According to the Canada's Energy Future 2013 report, natural gas resources are large enough to meet Canadian energy needs for many decades. In the specific case of this type of energy source, natural gas-fired power generation capacity is expected to increase substantially until 2035. On the other hand, natural gas production will probably grow 25% above current levels by 2035, led by higher levels of tight and shale gas development. Canadian electricity supply is also likely to increase by 27% during the coming years.

According to the Canada International Data and Analysis (2015) report, Canada is one of the "world's largest producers of dry natural gas[16] and the source of most US natural gas imports. Despite holding a smaller share of the world's proved natural gas reserves relative to crude oil, Canada ranked fifth in dry natural gas production and is a net exporter of dry natural gas". At the same time, "Canada is the fourth-largest exporter of natural gas, behind Russia, Qatar, and Norway. Although Canada has plans to export LNG in the future, all of Canada's current natural gas exports are sent to US markets via pipelines" (Canada International Data and Analysis, 2015).

Natural Gas Reserves in Canada

Canada has a total proved natural gas reserves, at the end of 2017, of 66,5 trillion cubic feet, which represents 1% of the world total natural gas reserves (6.831,7 trillion cubic feet in 2017).[17] Canadian proved natural gas reserves represent 17,7% of the total proved natural gas reserves in the North America region (excluding Mexico), estimated to be 375 trillion cubic feet. By its proved natural gas reserves, in 2017, Canada occupied the 15th position at the world level. The ratio/production is 10,7, which is lower than the US ratio/production (11,9) (BP Statistical Review of World Energy 2018, 2018).

Significant unconventional gas resources in the form of coalbed methane, shale gas, and tight gas are located within the Western Canada Sedimentary Basin. Canada has an estimated 573 trillion cubic feet of unproved technically recoverable shale gas resources, according to an updated assessment prepared by EIA. Five large sedimentary basins in western

---

[16] According to the Energy Markets Fact Book 2014—2015 report, Canada was, in 2014, the fifth natural gas producer and the fourth natural gas exporter, mainly to the US energy market. On the other hand, and considering the content of the Natural Gas Facts (2018) report, the country "natural gas reserves are estimated between 864 and 1. 773 trillion cubic feet, 281 to 323 trillion cubic feet of which is conventional gas, the rest is unconventional gas, including coal-bed methane, shale, and tight gas".

[17] In 2014, according to EIA sources, Canada had an estimate of proved natural gas reserves of 67 trillion cubic feet. The increase of proved natural gas reserves from 2014 to 2016 was 9,7 trillion cubic feet or 14,4% with respect to 2014. The marketable Canadian natural gas resources were reported to be 1.087 trillion cubic feet. On the other hand, marketable tight natural gas resources in Canada are estimated to be 528 trillion cubic feet. This includes 447 trillion cubic feet in the Montney play, located in northeastern British Columbia and northwestern Albert. Finally, the total Canadian shale resource is estimated to be 222 trillion cubic feet (Canada's Energy Future 2016, 2016).

Canada with thick, organic-rich shales—the Horn River, Cordova Embayment, and Liard in northern British Columbia; the Deep Basins in Alberta and British Columbia; and the Colorado Group in central and southern Alberta—"account for 536 trillion cubic feet of the total of technically recoverable shale gas resources" (Canada: EIA Country Overview, 2014). "The Liard Basin claims the largest share of the total resource. The remaining assessed resources are in the potential shale gas plays of Saskatchewan/Manitoba, Quebec, and Nova Scotia" (Canada International Data and Analysis, 2015).

Canadian natural gas supplies will still last many decades even if all proposals for new tanker shipments to Asia succeed. As a result of the latest estimates on natural gas proved reserves in Canada prepared by the National Energy Board and the US EIA, there is a vast abundance of this specific type of fossil fuel to serve Canadian needs for many decades.

The evolution of the natural gas proved reserves in Canada during the period 2010—17 is shown in Fig. 3.23.

According to Fig. 3.23, the Canadian natural gas proved reserves decreased by 4,8% during the period 2010—17, falling from 69,8 trillion cubic feet in 2010 to 66.5 trillion cubic feet in 2017. The peak in the level of natural gas proved reserves was reported in 2015. After that year, the level of natural gas proved reserves dropped each year. However, it is anticipated that the natural gas proved reserves in Canada, especially of tight and shale gas, will continue to increase during the coming years if new deposits are found as projected.

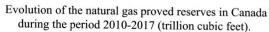

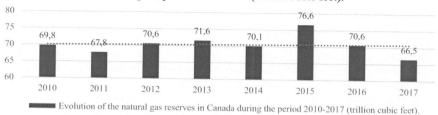

Evolution of the natural gas proved reserves in Canada during the period 2010-2017 (trillion cubic feet).

Figure 3.23 Evolution of the natural gas proved reserves in Canada during the period 2010—17. *(Source: BP Statistical Review of World Energy 2017 and BP Statistical Review of World Energy 2018 (June 2018).)*

Natural Gas Production in Canada

Natural gas has been part of Canada's energy mix since 1859 when it was first discovered in New Brunswick and later in Ontario in 1866, Alberta in 1883, and offshore Nova Scotia in 1967 (What is Natural Gas?, 2018). Canada natural gas resources are located in several provinces of the country. These provinces include: British Columbia, Alberta, Saskatchewan, Ontario, Quebec, New Brunswick, Nova Scotia, Prince Edward Island, Newfoundland and Labrador, the Northwest Territories, and Yukon. However, natural gas production is mainly from the following three provinces: British Columbia (Western Canada Sedimentary Basin), Alberta, and Saskatchewan. Atlantic Canada is currently the only region producing natural gas offshore, from areas off the coast of Nova Scotia (Natural Gas Facts, 2018). In the case of shale gas,[18] this type of natural gas is already being exploited in British Columbia and Alberta; substantial recoverable reserves may exist in Quebec, New Brunswick, Nova Scotia, and elsewhere in Canada.

During the period 2014–18, Canadian commercial natural gas production is expected to increase slightly (4%) rising from 14,7 billion cubic feet per day in 2014 to 15,4 billion cubic feet per day in 2018. However, after 2018, it is expected that the production of natural gas decreases by 5,2% to 14,6 billion cubic feet per day. After 2023, it is predictable that the production growth slowed and projected to be relatively stable after that, reaching 16,8 billion cubic feet per day by 2040, its highest level since 2007 (Canada's Energy Future 2017, 2017).[19]

The evolution of the natural gas production in Canada during the period 2012–17 is shown in Fig. 3.24.

Analyzing the data included in Fig. 3.24, the following can be stated: the production of natural gas in Canada during the period 2012–17 increased by 17,3% rising from 5.307,79 billion cubic feet in 2012 to 6.225,98 billion cubic feet in 2017. It is important to single out that the

---

[18] Just like conventional gas, most of the shale gas produced is thermogenic methane. It may contain small amounts of other gases such as ethane, butane, pentane, nitrogen, helium, and carbon dioxide, and impurities, as does conventional gas, but in Canada it is mostly sweet (i.e., it has no sulfur content). Like conventional gas, shale gas can also be wet (contains commercial amounts of natural gas liquids like ethane and butane) or dry (contains very little or no natural gas liquids). Thermogenic methane can be differentiated from biogenic methane through isotopic analysis (Council of Canadian Academies, 2014).

[19] It is important to highlight that future natural gas prices are a factor of uncertainty in all projection made on the production of natural gas in Canada for the coming years.

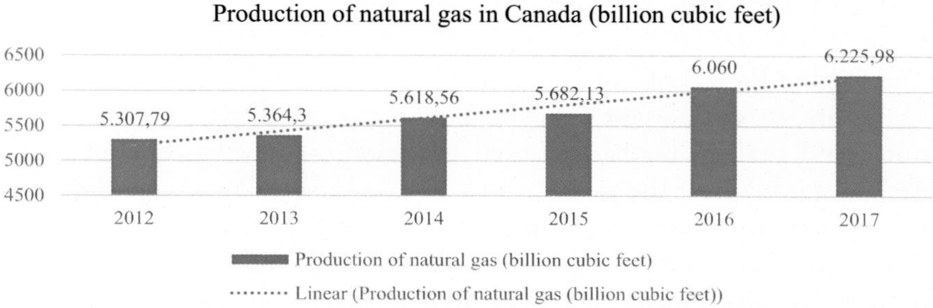

Production of natural gas in Canada (billion cubic feet)

Figure 3.24 Evolution of natural gas production in Canada during the period 2012—17. *(Source: BP Statistical Review of World Energy 2018 (June 2018).)*

production of natural gas increased each year within the period under consideration. The peak in the production of natural gas was achieved in 2017. It is likely that the output of natural gas in Canada could be around 13 billion cubic feet per day in 2018 (Magill, 2015). The production of natural gas in Canada represented, in 2017, about 4,8% of the world production of this type of energy source.

Recent technological advances in horizontal drilling and hydraulic fracturing have led to increased development of tight gas and shale gas resources in the Western Canadian Sedimentary Basin[20]. It is expected that this development will continue up to 2040, and, for this reason, it is projected that Canada's domestic natural gas production will grow to nearly 18 billion cubic feet per day by 2025; an expected increase of 29,2% concerning 2018 (Today in Energy, 2016).

Tight natural gas production in Montney formation, located in British Columbia and Alberta, reached in 2016, a total of 4,5 billion cubic feet per day (1.642,5 billion cubic feet), which represented 25% of the Canadian natural gas production for that year. It is expected that the Canadian tight gas production accounts for 70% of projected output in 2025, and that most of this production is likely to come from the Montney formation in British Columbia and Alberta, where tight gas production is projected to increase from 4,5 billion cubic feet per day in 2016 to 9,6 billion cubic feet per day in 2040.

The participation of conventional and non-conventional natural gas in the Canadian energy mix in 2040 is expected to be the following:
- conventional natural gas: 8% of the total;

---

[20] Natural gas production from the Western Canada Sedimentary Basin will increasingly come from shale gas, tight gas, and coalbed methane.

- tight natural gas: 76%; and
- shale gas: 6% (Canada's Energy Future 2016. Energy Supply and Demand Projections to 2040, 2016).

Finally, it is important to highlight that the production from conventional and coalbed methane natural gas resources in Canada is likely to decline steadily over the period up to 2040, as new drilling targeting these resources is not economic, given the natural gas price that is likely to be in that period. For this reason, western Canadian conventional non-tight production, which made up 37% in 2016, is expected to fall to 23% in 2040 (Canada's Energy Future 2017. Energy Supply and Demand Projections to 2040, 2017); this represents a reduction of 42% for the whole period considered. In the case of shale gas, Canada has been producing this type of natural gas since 2008, reaching 4,1 billion cubic feet in 2015.[21] Shale gas production in Canada is projected to continue increasing and to account for almost 30% of Canada's total natural gas production by 2040 (Shale Gas Production Drives World Natural Gas Production Growth, 2016).

In 2016, the Canadian production of conventional gas sources was 25% of the total. Within the unconventional gas, the output of tight gas represents the significant percentage in the joint production of tight and shale gas.

Dry Natural Gas Production in Canada

In 2012, the daily production of dry natural gas in Canada reached 13,89 billion cubic feet. This production allows Canada to occupy the fourth place at the world level in dry natural gas production, behind Russia, Iran, and Qatar. The evolution of the output of dry natural gas in Canada during the period 2012—15 is shown in Fig. 3.25.

According to Fig. 3.25, the production of dry natural gas increased by 4,4% during the whole period considered. However, it is important to highlight that the output of dry natural gas increased by 5,5% during the period 2012—14, but decreased by 1,1% during the period 2014—15. The peak in the dry natural gas production was achieved in 2014.

---

[21] In 2014, shale gas accounted for approximately 4% of total Canadian natural gas production while tight gas accounted for 47%. By 2035, the National Energy Board expects that tight and shale gas production together will represent 80% of Canada's natural gas production (Exploration and Production of Shale and Tight Resources, 2016). By 2040, tight and shale gas production is expected to increase to 84% of Canada's natural gas production.

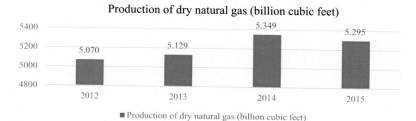

Figure 3.25 Evolution of the dry natural gas production in Canada during the period 2012—15. (*Source: EIA database 2018 and Statista database 2017.*)

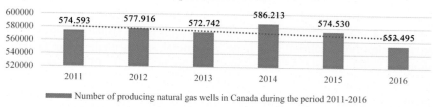

Figure 3.26 Evolution in the number of producing natural gas wells in Canada during the period 2011—16. (*Source: EIA database 2018.*)

On the other hand, there is a high number of wells in Canada associated with the production of natural gas. The evolution in the amount of producing wells in Canada during the period 2011—16 is shown in Fig. 3.26.

Fig. 3.26 shows that the number of producing wells in Canada decreased by 3,7% during the period 2011—16. The peak in the number of producing wells within the period considered was achieved in 2014. After that year the number of producing wells in Canada decreases by 5,6%. It is expected that this trend will continue during the coming years as a result of the increase in the production of natural gas in the US, and the drop of the imports of this specific type of energy source by this country.

Small-Scale Liquefaction Natural Gas and Potential for Fuel-Switching in Canada

Small-scale LNG facilities have been in operation in Canada in the provinces of Ontario, Quebec, and British Columbia for more than four decades. These facilities are used in conjunction with LNG storage facilities, with the aim of supplying natural gas when the demand for this type of energy source exceeds the capacity constraints of nearby pipelines. However, it is LNG's role as a substitute for diesel that may be a key driver of investments in small-scale LNG facilities into the future.

However, it is important to highlight that there is considerable uncertainty related to the adoption of LNG as a diesel alternative in Canada. Among the main difficulties in the use of LNG as a diesel alternative in the country are technological development, and the relatively low prices of natural gas and diesel. Gradual adoption of LNG as a diesel alternative in selected areas is expected to take place during the coming years, taking into account the government and industry projections.

In Canada, natural gas costs less than diesel for an equivalent amount of energy. This difference in cost has created interest in using LNG domestically as a diesel substitute. LNG for domestic use is usually produced by small-scale LNG facilities, which are, in general, much smaller than facilities used for LNG exports. A significant challenge in the use of LNG in Canada has been the lack of small-scale liquefaction infrastructure. Without nearby plants to liquefy natural gas, LNG must be shipped via truck over long distances, increasing the cost for users and reducing its competitiveness compared to other types of energy sources. However, the recent and proposed construction and expansion of small-scale LNG facilities will decrease shipping distances for many future LNG users (Canada's Energy Future 2016. Energy Supply and Demand Projections to 2040, 2016).

LNG can be considered as an effective alternative for the supply of natural gas to a given region, in those cases when the demand exceeds the installed capacity of the pipeline. It can be not only expensive to expand the size of the pipeline, especially if the additional capacity to be built is expected to be used only a few days a year, but it is very likely that it will not be profitable. Construction of LNG storage tanks and the production of specialized trucks for transportation can be a viable and less expensive alternative to be considered by the Canadian government. For this reason, there are multiple proposals to build LNG facilities on Canada's west and east coasts currently under consideration of Canada's National Energy Board, all aiming to export LNG to global markets. The final volume of LNG exports from Canada is significantly uncertain for the projections included in the Canada's Energy Future 2016 report. According to this report, LNG could also be used to convert existing remote distribution networks from propane to natural gas.

Natural Gas Consumption in Canada

During the period until 2035, the total energy consumed by Canadians is expected to continue to grow, but at a slower rate than in the past. Hydrocarbons are, and expected to continue to be, the primary source of energy to heat homes and in businesses, to transport people and goods, as well as in several other functions that are integral to the standard of living of the Canadian people (Canada's Energy Future 2013 — Energy Supply and Demand Projections to 2035, 2013).

The country demand for oil and natural gas is expected to increase by 28% over the period up to 2035, but this predicted increase in the consumption of natural gas will depend on its price. According to the Canada's Energy Future 2017 report, natural gas prices have declined considerably over the past decade falling from US$6-9 per million British thermal units' range between 2006 and 2008 to less than US$4 per million British thermal units for the majority of the last five years. In the first half of 2016, prices were as low as US$2 per million British thermal units and averaged closer to US$3 per million British thermal units in the second half of 2016 and into 2017. It is anticipated that the price of natural gas in the Canadian market will be around US$4,30 per million British thermal units in 2040.

It is well known that natural gas is the less contaminated energy source in comparison to other fossil fuels, such as oil or coal. However, it is unclear whether natural gas demand would increase or decrease in the coming years. The final trend will depend on the decision of energy producers in the North America region. If they decide to replace or convert coal–fired power plants into natural gas–fired power plants to reduce CO_2 emissions, then an increase in the consumption of this type of energy source will be registered during the coming years. If this happens, then a global shift toward natural gas consumption could also increase demand for LNG exports from North America, which could raise natural gas prices. On the other hand, the adoption of stronger climate policy by Canada could reduce the demand for all fossil fuels, including natural gas, during the coming years.

The Canadian government is adopting a group of measures to encourage the use of different energy sources available in the country, including natural gas, in the industrial sector,[22] which uses natural gas for

---

[22] According to the International Energy Outlook 2016 report, in Canada natural gas consumption in the industrial sector is expected to grow by an average of 0,2% per year during the period 2012—20 and by 1,1% per year during the period 2020—40 (an average increase of more than five times).

various purposes, such as for heat source, a fuel for the generation of steam, and a feedstock in the production of petrochemicals and fertilizers. The electric power generation sector uses natural gas as a fuel for electricity generation and heating. In addition to the use of natural gas in the industry, natural gas is also used extensively in residential, commercial, and power generation applications. In the case of residential and commercial users, natural gas is mostly used as a source of space heating, water heating, clothes drying, and in cooking applications.

Due to the increased use of natural gas for electricity generation and heating in Canada in the last years, it is crucial to adopt additional measures to improve energy efficiency. By 2035, the energy used per unit of economic output is projected to be 20% lower than the level reported in 2012, due to improvements in energy efficiency. In a reversal of the long-term trend, passenger transportation energy use is expected to decline over the projected period, mainly due to new passenger vehicle emission standards, which are expected to improve vehicle fuel efficiency (Canada's Energy Future 2013. Energy supply and Demand Projections to 2035, 2013).

The evolution of natural gas consumption in Canada during the period 2010—17 is shown in Fig. 3.27.

According to Fig. 3.27, the consumption of natural gas in Canada during the period 2010—17 increased by 30,4% rising from 3.132,41 billion cubic feet in 2010 to 4.085,9 billion cubic feet in 2017. The peak in the use of natural gas within the period considered was reached in 2017. It is projected that the consumption of natural gas will increase at an average of 1,1% during the period 2020—40, due to a reduction in the participation of coal in the country energy mix, as a result of the closure of a substantial number of old and ineffective coal-fired power plants.

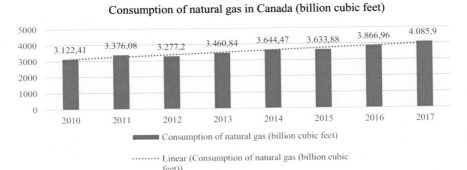

Figure 3.27 Evolution of natural gas consumption in Canada during the period 2010—17. (Source: BP Statistical Review of World Energy 2018 (June 2018).)

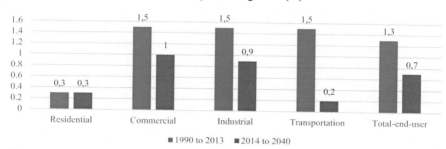

Average annual growth (%)

Figure 3.28 Past and future growth in end-use energy demand by sector. *(Source: Canada's Energy Future 2016 report.)*

From Fig. 3.28, the following can be stated: it is evident that during the period 1990–2013, the average annual growth was much higher than the growth projected for the period 2014–40. The above figure also shows that the energy use increased by 1,3% per year as an average during the period 1990–2013. For the planned period up to 2040, energy use is expected to grow nearly half the past period considered (0,7%). The total-end-user projected for the period 2014–40 is 46,2% lower than the growth registered during the period 1990–2013. The most significant reduction is recorded in the transportation sector (87%), expected to reach the lowest increase considering all five areas included in Fig. 3.28.[23]

Finally, it is important to highlight that the demand for natural gas in Canada varies depending on several factors such as weather, economic growth, market conditions, and infrastructure constraint. The request of natural gas in the country is also driven by the industrial sector (43%), followed by the electric generation sector (22%), the residential sector (20%), and the commercial sector (16%). In 2017, the production of natural gas in Canada exceeds the consumption of this type of energy source, increasing the opportunity for the exports of natural gas to the US and other countries as well (Canada's Natural Gas, 2018).

[23] Natural gas use in the transportation sector is increasing and it is expected to continue growing in the coming years. "Natural gas vehicles (NGVs) use either compressed natural gas or LNG. In the longer term, the projections include a moderate penetration of NGVs in both forms. The outlook also accounts for the recent adoption of LNG use by ferries, and assumes moderate levels of LNG adoption by marine tankers and rail locomotives" (Canada's Energy Future 2016, 2016).

Natural Gas Consumption in Oil Sands in Canada

Oil sands extraction is an activity that consumes a lot of energy and, in the case of Canada, consumes a considerable amount of natural gas as fuel. Over the last several decades, the extraction of oil sands has been a significant growth area for natural gas consumption in Canada. Oil sands natural gas demand, including gas consumed for cogeneration, increased from 0,7 billion cubic feet per day in 2000 to 3,1 billion cubic feet per day in 2014, an increase of 4,4 times. This demand now represents over 20% of the total Canadian natural gas demand (Alberta's Energy Reserves and Supply/Demand Outlook, 2016).

However, due to continuous technological improvements that are found and incorporated into the oil sands extraction process, the amount of natural gas needed to produce a bitumen barrel is expected to decrease 1% during the coming years. On the other hand, during the period 2015—25, it is projected that the intensity of the oil sands will increase slowly. After 2025, efficiency improvements are expected "to begin to outweigh the effect of growing in situ production, resulting in declining natural gas intensity for the remainder of the projection period. Natural gas requirements for oil sands use, including those for electrical cogeneration is expected to rise 9,7% from 3,1 billion cubic feet per day in 2014 to 3,4 billion cubic feet per day by 2040" (Canada's Energy Future 2016, Energy Supply and Demand Projections to 2040, 2016).

In addition to the use of natural gas for oil sands extraction, this type of energy fuel is also used in mining, upgrading, and in situ oil sands production. This last activity can be broadly categorized into two categories: (a) primary and (b) thermal. During primary production, natural gas has become a smaller component of the total in situ production. It is expected that the use of natural gas in total in situ production will decline from 42% in 2000 to 11% by 2040; this represents a reduction of 31%. In thermal in situ process, "natural gas is used to heat water to produce steam, which is injected into the ground to heat the bitumen in the oil reservoir. Heating the bitumen reduces its viscosity, allowing it to flow to the well and be pumped to the surface" (Canada's Energy Future 2016- Energy Supply and Demand Projection to 2040, 2016).

Exports and Imports of Natural Gas in Canada

The US is the only country in the world to which Canada exports natural gas. Total exports of natural gas from Canada to the US declined during the period 2007—14 from 10,4 billion cubic feet per day in 2007 to 7,4 billion cubic feet per day in 2014; this represents a decrease of 29% during the whole period. The US midwest typically receives about half of Canada's natural gas exports, with the remainder split between the other two regions: eastern and western. The volume of natural gas exports to the west region of the US has remained relatively stable during the last years, while natural gas export volumes to the US eastern and midwest regions have declined, mainly due to an increase of natural gas production in the US (Canada's Energy Future 2016-Energy Supply and Demand Projections to 2040, 2016). In 2017, Canada exported 8,2 billion cubic feet per day of natural gas to the US (Natural Gas Facts, 2018).

According to the report mentioned above, it is projected that during the coming years net exports of natural gas from Canada to the US will continue to decline until 2019 when they begin increasing with the start–up of LNG exports. After 2023, total net exports of natural gas by Canada are expected to decrease to 3,7 billion cubic feet in 2040.

Despite the expected drop of Canada exports of natural gas to the US in the future, Canada will continue to be a net exporter of natural gas to the US, with growth in LNG export volumes replacing some of the export volumes lost throughout the pipeline. It is expected that, by 2040, Canada's net exports of natural gas will be 22% higher than in 2012 (International Energy Outlook, 2016), and that Canadian imports as a percentage of consumption of natural gas will increase, reaching just over 10% in 2035, due, among other factors, to the increase of Mexican imports of natural gas.

The evolution of the imports and exports of natural gas and LNG by Canada during the period 2011—17 is included in Table 3.8.

According to Table 3.8, natural gas and LNG exports by Canada decreased by 9,4% falling from 8,7 billion cubic feet per day in 2011 to 8,2 billion cubic feet per day in 2017. Pipelines to the US carried out most Canadian natural gas exports. Although 2016 saw an increase of 16,2% in

Table 3.8 Evolution of the Imports and Exports of Natural Gas and LNG by Canada During the Period 2011—17 (billion cubic feet per day).

Volumes	2011	2012	2013	2014	2015	2016	2017
Exports	8,7	8,4	7,8	7,4	7,4	7,9	8,2
Imports	3	3	2,6	2,1	1,9	2,1	2,4
Net exports	5,7	5,4	5,3	5,3	5,5	5,8	5,8

Source: Canadian National Energy Board, BP Statistical Review of World Energy 2017 (2017) and Natural Gas Facts, 2018. Natural Resources Canada.

natural gas and LNG exports from Canada to the US, there has been a steady decline in natural gas and LNG exports since 2011, except for 2016 and 2017. This decline is the result of an increase in the US domestic supply of this type of energy source. The US is now a net exporter of natural gas for the first time in almost six decades. At the same time, during the period considered there has been a continued modest delivery of LNG and CNG exported by truck from Canada to the US.

In the specific case of natural gas imports, the following can be stated: the volume of natural gas imports by Canada during the period 2011–17 decreased by 20% falling from 3 billion cubic feet per day in 2011 to 2,4 billion cubic feet per day in 2017. The imports of natural gas by Canada declined during the period 2011–15 but began to increase during the period 2015–17.

In summary, the net exports trade from Canada to the US during the period under consideration increased by 1,8% rising from 5,7 billion cubic feet in 2011 to 5,8 billion cubic feet in 2017. It is expected that the natural gas trade between Canada and the US will continue this trend at least during the coming years. The value of Canadian net exports reached more than US$7 billion in 2015. In 2017, the amount of Canadian net exports of natural gas dropped by 4,3% and reached US$6,7 billion.

The evolution of the natural gas trade between Canada and the US during the period 2011–17 is shown in Fig. 3.29.

As can be easily seen in Fig. 3.29, Canada exports not only natural gas to the US but also imports natural gas from this country, mostly into Ontario. Small amounts of natural gas received as LNG are also imported from the US. Total natural gas imports by Canada increased more than doubled during the period 2007–2011, rising from 1,3 billion cubic feet per day in 2007 to 3 billion cubic feet per day in 2011. However, the imports of natural gas by Canada from the US after 2011 began to decline gradually in

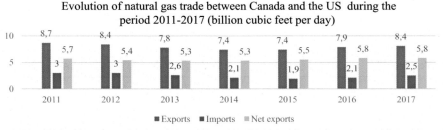

Figure 3.29 Evolution of natural gas trade between Canada and the US during the period 2011–17. (*Source: Table 3.8.*)

subsequent years until 2015. Since that year the imports of natural gas by Canada from the US began to increase once again rising from 1,9 billion cubic feet per day in 2015 to 2,4 billion cubic feet per day in 2017. In the case of the exports of natural gas, Fig. 3.29 shows a decrease of 15% during the period 2011—15 and an increase of 10,8% during the period 2015—17.

Summing up the following can be stated: according to EIA sources, bilateral natural gas trade between Canada and the US is dominated by pipeline shipments. "Natural gas imports from Canada increased to 8,1 billion cubic feet per day in 2017, accounting for 97% of all US natural gas imports. Total natural gas imports from Canada were valued at US$7,3 billion in 2017" (The US is a Net Energy Importer from Canada, 2018). On the other hand, "US natural gas exports to Canada, which increased to 2,5 billion cubic feet per day in 2017, mainly go from New York into the Eastern Canadian provinces". Increases in pipeline capacity to carry natural gas out of the Marcellus and Utica shale formations increased flows of US natural gas into Canada, reducing pipeline imports from this country, and rising US pipeline exports to Canada (The US is a Net Energy Importer from Canada, 2018).

Electricity Generation Using Natural Gas as Fuel in Canada

The structure of the Canadian electricity generation sector has been changed significantly over the last 10 years. In most provinces, there has been a shift from vertically integrated operating as regulated monopolies to electric utilities operating to different degrees of market liberalization and unbundling of electricity generation, transmission, and distribution services (About Electricity, 2016). The Canadian industry accounts for the largest share of the electricity generated or available in the country, as a result of the presence of several energy-intensive industrial activities.

In Canada, natural gas-fired power capacity accounted for 15% of the total capacity installed in the country in 2016 (21,5 GW installed), considering all available energy sources.

It is important to single out that the government is adopting an energy policy with the aim of reducing the emissions of contaminated gases to the atmosphere and, for this reason, natural gas is a preferred energy source for electricity generation and heating. As a result of the implementation of this energy policy, natural gas-fired power plants will replace several old and ineffective coal and oil-fired power plants during the coming years, increasing its participation in the energy mix of the country. Due to these substitutions in the Canadian electricity generation sector, it is expected that natural gas-fired power plants capacity will grow steadily up to 2040. Based on this assumption, it is projected that natural gas-fired power capacity increases almost 94% rising from 21,5 GW in 2016 to 41,7 GW in 2040.

An essential element to single out is that all regions in Canada other than Prince Edward Island and Nunavut use natural gas for electricity generation and heating. The two areas that generate the most significant amount of electricity using natural gas as fuel are Alberta and Saskatchewan.

As a result of relatively low fuel prices and upfront capital costs, natural gas-fired power capacity is expected to replace much of the old and ineffective coal-fired power capacity that will be retired up to 2040. Natural gas-fired power capacity is also expected to grow as a result of increasing use of different renewable energy sources such as wind and solar energies for electricity generation and heating in the country. Higher production of electricity by renewable energy sources "requires grid operators to be able to accommodate fluctuations in both electricity consumption and production. This makes natural gas power plants' ability to increase or decrease generation quickly an attractive quality, especially in provinces without large-scale hydro resources" (Canada's Energy Future 2017. Energy Supply and Demand Projections to 2040, 2017).

It is also important to highlight that Canada only exports electricity to the US. Over 60% of the electricity exports by Canada are sold into the eastern half of the US, the vast majority of which is sold to the US northeast. States in the US midwest receive about one-quarter of Canada's electricity exports, while about 13% is exported to the US west coast. The electricity exports by Canada to the US are predominately generated in provinces with substantial hydroelectric generation capability installed such as Quebec, British Columbia, and Manitoba.

The evolution of electricity generation and heating in Canada using natural gas as fuel during the period 2010—17 is shown in Fig. 3.30.

According to Fig. 3.30, the electricity generation and heating in Canada using natural gas as fuel during the period 2010—17 increased by 53,6% rising from 47,81 TWh in 2010 to 73,4 TWh in 2017. The peak in the electricity generation and heating during the period considered was reached in 2017. In this year, the total electricity generated in Canada using all available energy sources reached the total of 693,4 TWh. In 2017, the power produced using natural gas as fuel represented 10,6% of the total. It is expected that the use of natural gas for the electricity generation and heating in the country will continue to increase during the coming years, as a result of the closure of several old and ineffective coal-fired power plants.

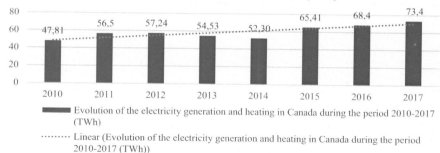

Figure 3.30 The evolution of electricity generation in Canada using natural gas as fuel during the period 2010—17. *(Source: BP Statistical Review of World Energy 2018 (June 2018).)*

It is likely that the role of natural gas in the energy mix of Canada in the future will be higher than today.

It is foreseen that, in 2035, the total electricity generated by natural gas-fired power plants will reach 114.473 GWh; this represents an increase of 80,8% with respect to the level registered in 2016. However, hydropower plants have been, and will continue to be, the dominant energy source for electricity generation and heating in Canada during the coming decades. In 2016, hydropower plants generated 379.630 GWh or 62% of the total electricity produced in the country in that year.

The adoption in August 2015 of the CPP by the US government sets emission reduction goals for 47 US states. The US government to meet their emission reduction targets should import electricity from Canadian sources creating a new market opportunity for Canadian electricity exports.

Summing up the following can be stated: electricity accounts for a small but locally significant share of bilateral trade. In 2017, the value of US imports of electricity from Canada increased for the second straight year, reaching US$2,3 billion. The US imported 72 million MWh of electricity from Canada in 2017, and exported 9,9 million MWh, based on data from the Canadian government.

CHAPTER 4

Current Status and Perspective in the Use of Coal for Electricity Generation in the North America Region

Contents

General Overview

Coal plays a vital role in the security of energy supply in developed countries and is a crucial factor for economic growth and development in

Conventional Energy in North America
ISBN 978-0-12-814889-1
https://doi.org/10.1016/B978-0-12-814889-1.00004-8

many developing countries. Coal resources exist in many countries, including those with significant energy challenges, and are used for electricity generation and heating in many of them. Therefore, coal has a vital role to play in assisting the development of baseload electricity which is most needed in these countries (World Energy Resources Coal 2016, 2016), without taking into account the kind of country that needs it, and despite the negative consequences for the environment and population.[1]

In the era of modern power plants, coal has always generated more electricity in the US than any other fuel source. In recent decades, other sources have been competing for the second place: first hydroelectricity, then natural gas, nuclear power, and natural gas again. In the case of Canada, the primary energy source for electricity generation and heating is hydropower with 62% of the total electricity generated by the country in 2016.

Many developing countries are increasingly satisfying their growing energy demands with cheap coal to sustain economic growth, to reduce energy poverty, and to achieve the United Nations development goals. Many countries in Asia and Africa are currently making significant investments in new coal infrastructures albeit with clean coal technologies (Mercator Research Institute on Global Commons and Climate Change, 2015).

The incremental demand for coal is visible in some regions, notably in the non–OECD Asia countries, because they are focused on maintaining the potential for continued economic growth. But, at the same time, these

---

[1] Coal is the dirtiest of all conventional energy sources available in the world (oil, natural gas, and coal). The use of coal for electricity generation and heating releases around 60 hazardous toxic pollutants into the air caused by CO_2 emissions, water by poisoning of local rivers, by acid mine drainage, and land. Some of these toxic pollutants cause cancer, others cause damage to the nervous and immune systems, while some others impede reproduction and affect development. The main substance that are produced by coal-fired power plants are: sulfur dioxide, nitrogen oxides, hydrogen chloride, hydrogen fluoride, arsenic, cadmium, chromium, mercury, and dioxin. Pollution from coal-fired power plants "is released in four main ways; (1) as fly ash from the smoke stack, (2) bottom ash which stays at the bottom after the coal is burned, (3) waste gases from the scrubber units (which are chemical processes used to remove some pollutants), and (4) gas released into the air (The Environmental Impacts of Coal, 2005). According to the BP Statistical Review of World Energy 2018 report, world proved coal reserves are currently sufficient to meet 134 years of global production, much higher than the R/P ratio for oil and gas.

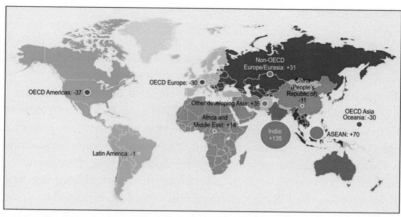

Figure 4.1 IEA (2017), Coal 2017: Analysis and Forecasts to 2022, IEA, Paris, https://doi. org/10.1787/coal_mar-2017-en. The figure is Map 3.1 Incremental global coal demand 2016-22 (Mtce), page 70 under the Global coal demand forecast. Note: This map is without prejudice to the status of or sovereignty over any territory, to the delimitation of international frontiers and boundaries and the name of any territory, city or area.

countries are adopting measures for the protection of the environment from excessive accumulation of anthropogenic greenhouse gas (GHG) emissions and other air pollutants (see Fig. 4.1).

From Fig. 4.1, the following can be stated: the countries and regions where the use of coal for electricity generation and heating is expected to grow during the period 2016—22 are India, Europe/Eurasia, ASEAN countries, and other developing countries in Asia, Africa, and the Middle East. On the other hand, the countries and regions where the use of coal for electricity generation and heating is projected to decrease are OECD Americas, OECD Europe, China, and OECD Oceania.

According to the World Energy Resources 2016 report, coal is known as the most carbon-intensive fossil fuel available in many countries from different regions. For this reason, the use of coal for electricity generation and heating could have negative implications for climate change mitigation strategies adopted by many countries. However, it is important to highlight that the negative impact on the environment and population through the use of coal for electricity generation and heating can be significantly reduced if low emissions and high-efficiency technology are utilized in high proportions.[2] "With modern technological advancements, coal-fired power plants could follow technologies that allow higher efficiency and low carbon emissions to tackle climatic changes" (World Energy Resources, 2016 used with the permission of the World Energy Council www.worldenergy.org, 2016).

In the North America region, most of the coal is consumed in the US, which accounted for almost 93% of the region's total coal use in 2016. It is expected that the use of coal in the US for electricity generation and heating will remain relatively flat until 2040 rising by only 2 quadrillions Btu. However, if the proposed CPP were implemented, then the US coal consumption is expected to decline by almost three quadrillions Btu by that year. In this case, the US coal consumption would be expected to decrease by around 25% in 2040 concerning the level of consumption registered in 2016. Moreover, strong growth in shale gas production, slowing electricity demand as a result of the implementation of energy efficiency measures, the adoption of environmental regulations to reduce the negative impact of the use of some fossil fuel for electricity generation and heating, and the increased use of renewable energy sources for electricity generation and heating, is expected to reduce the share of coal-fired generation within the total US electricity production (including electricity generated at plants in the industrial and commercial sectors) from 37% in 2012 to 26% in 2040; this means a decrease of 9% for the whole period (International Energy Outlook 2016 with Projections to 2040, 2016).

The main factors impacting coal consumption in the US are shown in Table 4.1.

In 2017, according to the BP Statistical Review of World Energy 2018 report, the primary energy consumption of coal in the North America region decreased by 2,5% with respect to the level registered in 2016, reaching 350,7 million tons of oil equivalent or 9,4% of the world total (3.731,5 million tons of oil equivalent). In 2016, the primary energy

---

[2] Restrictions on coal production and in the use of coal for electricity generation and heating in the US have reached unprecedented severity under the former Obama administration. However, the Trump administration is likely to rescind many of the restrictions imposed by the former Obama administration on the use of coal for electricity generation and heating, such as a new slate of restrictions announced in 2017. This may not revive the coal industry, which faces strong competition from inexpensive natural gas. Nevertheless, it is expected that the use of coal for electricity generation and heating will face fewer regulatory restrictions under the Trump administration (Taylor, 2016).

Table 4.1 Factors Impacting Coal Consumption in the US.

Country	Climate Policy	Economic Growth	Gas Competition
United States	A negative factor for coal consumption	A neutral factor for coal consumption	A negative factor for coal consumption

Source: World Energy Outlook 2015.

consumption of coal in the North America region represented 9,7% of the world total (3.706 million tons of oil equivalent). By regions, North America is the one with the lowest consumption of coal in the world.

In 2017, the US coal production reached 371,3 million tons of oil equivalent, an increase of 6,6% concerning the level registered in 2016 (348,3 million tons of oil equivalent), which is the lowest annual coal production since 1979 (BP Statistical Review of World Energy 2018, 2018). Out of this total, 45,3% was subbituminous (used primarily for electricity generation), 44,4% bituminous (mainly used for electricity generation, heat, and power applications), 10% lignite (also known by brown coal and is used almost exclusively for electricity generation), and 0,2% anthracite (used for home heating and steel making). In the same year, the US coal consumption was 332,1 million tons of oil equivalent, which is 2,4% lower than the level registered in 2016 (BP Statistical Review of World Energy 2018, 2018). "The electric power sector accounted for about 92,8% of the total US coal consumption in 2016" (Annual Coal Report, 2017).

In 2017, Canada's coal production reached 31,1 million tons of oil equivalent, representing a decrease of 2,3% concerning 2016. The coal consumption in Canada, in 2017, reached 18,6 million tons of oil equivalent, representing a decrease of 1,6% concerning the level registered in 2016 (BP Statistical Review of World Energy 2018, 2018).

It is important to note that the largest coal producing countries are not confined to only one region. The top six coal producers are China, US, India, Indonesia, Australia, and South Africa. Europe is the only region without a significant coal producer. "Much of global coal production is used in the country in which it is produced, and only around 18% of hard coal production is destined for the international coal market" (World Energy Resources, 2016 Coal used with the permission of the World Energy Council www.worldenergy.org, 2016).

However, and despite the decrease in the world coal production in the last three years falling from 4.006,1 million tons of oil equivalent in 2013 to 3.656,4 million tons of oil equivalent in 2016, coal still provides around

40% of the world's electricity. Climate change mitigation demands, the transition to cleaner energy forms, and increased competition from other energy sources are presenting serious challenges for the coal sector at the global level. Asia offers the biggest market for coal and in 2016 accounted for 73,8% of global coal consumption (World Energy Resources 2016 Coal Coal, used with the permission of the World Energy Council www.worldenergy.org, 2016). This situation is not expected to change any time soon.

According to the International Energy Outlook 2016 report, world coal production is expected to increase by 11,1% rising from 9 billion tons in 2012 to 10 billion tons in 2040, with much of the growth occurring outside the North America region. China, the world's leading coal producer, is expected to decrease its share in the consumption of coal from 48% in 2012 to 44% in 2040; this means a decrease of 4% during the period considered. It is important to highlight that the world coal production varies significantly from region to region with an expected sustained strong growth in India, slowing growth and a gradual decline in China, particularly after 2025, and little change in the US and OECD Europe.

Consumption of steam coal is projected to grow by 20% from 2013 to 2040 (World Energy Resources 2016 Coal, used with the permission of the World Energy Council www.worldenergy.org, 2016). Lignite, also used in electricity generation and heating, has been forecasted to rise until 2020.

Without a doubt, "coal is playing and will continue to play a key role in the world's energy mix, with demand in certain regions set to grow rapidly. Growth in both the steam and coking coal markets will be strongest in developing Asian countries, where the demand for electricity and the need for steel in construction and car production, and demands for household appliances will increase as incomes rise" (World Energy Resources 2016, used with the permission of the World Energy Council, www.worldenergy.org, 2016).

With no global upturn in coal demand in sight due, among other elements, to environmental considerations, the search for market equilibrium depends on cuts to supply capacity, mainly in countries with a high coal demand for electricity generation and heating, such as China and the US. There are stark regional contrasts in the coal demand projections for the coming decades. Some higher-income economies, most of them within the group of developed countries, often with flat or declining overall energy needs as a result of the introduction of significant energy efficiency measures, are taking additional steps to reduce further the use of coal for electricity generation and heating (World Energy Resources 2016 Executive Summary, used with the permission of the World Energy Council www.worldenergy.org, 2016). These countries are searching for lower carbon alternatives to change their energy mix, and are increasing the use of natural gas and renewable energy sources, primarily solar and wind energies, for electricity generation and heating. Coal demand in the EU and the US (which together account for around one-sixth of today's global coal use) is expected to reduce the use of coal for electricity generation and heating by over 60% and 40% respectively, over the period up to 2040.

In Canada, there are two main types of coal produced: (1) thermal and (2) metallurgical. Canadian thermal coal production is linked to the use of coal in the electricity sector, particularly in Alberta and Saskatchewan provinces. On the other hand, metallurgical coal is primarily used for steel manufacturing domestically and internationally. Much of Canada's metallurgical coal production is exported. It is important to single out that the Canadian future coal production trends are closely linked to the global coal demand and prices (Canada's Energy Future 2017. Energy Supply and Demand Projections to 2040, 2017).

In 2015, thermal coal demand for electricity generation and heating accounted for 93% of the total coal consumption in Canada. During the period 2016—40, the Canadian thermal coal consumption is expected to decline by 86% falling from 37,7 million tons in 2016 to 5,4 million tons in 2040. This decline more than offsets the increasing demand for coal in the industrial sector during the period under consideration. The expected decrease in the consumption of thermal coal during the projected period is due to the retirements of old and inefficient coal-fired power plants by 2030, efficiency improvements from retrofits, and the construction of new coal-fired power plants with CCS technology (Canada's Energy Future 2016, Energy Supply and Demand Projections to 2040, 2016).

Based on the information included in the above-mentioned report, the overall Canadian demand for coal is expected to decrease by 78% falling from 36,3 million tons in 2016 to 7,9 million tons in 2040. As a result of a projected 50% decrease in the consumption of thermal coal, its production in Canada is expected to decrease from 38 million tons in 2014 to 19 million tons by 2040; this means a decreased of 50%.

In Canada demand for metallurgical coal used in steel manufacturing is expected to remain stable at 3 million tons up to 2040. However, global demand for metallurgical coal is probable to grow moderately over the projection period, resulting in an expected steady growth in net exports of this type of coal from Canada during that period. Total metallurgical coal production in Canada is expected to increase from 26,9 million tons in 2016 to 30,5 million tons in 2040; this represents an increase of 13,4% (Canada's Energy Future 2017, Energy Supply and Demand Projections to 2040, 2017).

The overall demand for coal imports is expected to be nearly the same in 2040 as in 2013, with total coal imports to Asia slightly higher in 2040; to Europe/other countries (including Eurasia, the Middle East, and Africa) marginally lower; and in the North America region, in 2040, about the same as in 2013, according to the information included in the International Energy Outlook 2016 with Projections to 2040 report (see Figs. 4.2—4.5).

On the other hand, consumption of steam coal is projected to grow by 20% from 2013 to 2040. Lignite, also used in power generation, has been forecasted to increase during 2020. Demand for coking coal used in iron and steel production has more than doubled since 2000, but according to the World Energy Outlook 2015 report, demand will moderate over the coming decade as China enters a new phase of economic development. Several measures are already adopted by the Chinese government to reduce the use of coal for electricity generation and heating, including the most pollutant coal-fired power plants.

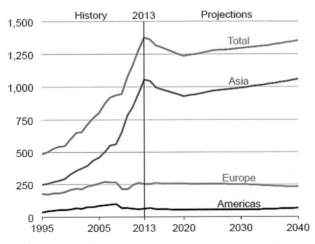

Figure 4.2 World coal imports by major importing region during the period 1995—2040 (million short tons). (*Source: International Energy Outlook 2016 with Projections to 2040.*)

Figure 4.3 Castle gate coal-fired power plant. *(Source: Courtesy David Jolley Wikipedia Commons.)*

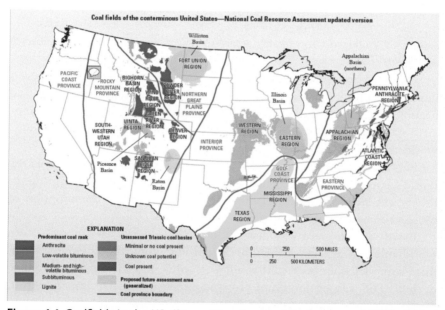

Figure 4.4 Coalfields in the US. *(Source: Department of the Interior and US Geological Survey database.)*

Without a doubt, the biggest market for coal trade during the period until 2040 is Asia, which currently accounts for 66% of the global coal consumption, although China is responsible for a substantial proportion of this level of use. Many countries do not have sufficient fossil fuel resources to cover their energy needs and, therefore, need to import energy to help meet their requirements. Japan, Chinese Taipei, and Korea, for example,

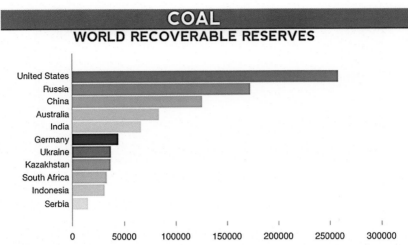

Figure 4.5 World coal recoverable reserves. *(Source: Institute for Energy Research database (2018).)*

import significant quantities of steam coal for electricity generation and heating and coking coal for steel production (World Energy Resources, 2016 Coal, used with the permission of the World Energy Council www. worldenergy.org, 2016).

It is anticipated that coal will continue to play a crucial role in the world's energy mix, with demand in specific regions set to overgrow. In others, the need for coal will drop significantly, due to the adoption of more strict environment measures.

Based on the relatively flat outlook for world coal imports, both worldwide and in each of the three major coal-importing regions, exports from some areas are expected to increase while in others are expected to decline. The lack of significant growth in total world coal imports represents a substantial change to the long-term historical trend of continuous annual growth, which led to significant increases in coal exports for some regions (International Energy Outlook 2016, 2016).

According to the International Energy Outlook 2016 with Projections to 2040 report, regions whose coal exports are expected to increase during the period considered include Australia (85 million short tons), South Africa (23 million short tons), Eurasia (19 million short tons), and South America (27 million short tons). On the other hand, it is expected that the exports of coal will decline, during the period 2013—40, in the following countries:

Indonesia (−95 million short tons), the US (−59 million short tons), and Vietnam and North Korea (−28 million short tons) each.[3]

It is important to note that most of the world's coal trade consists of steam coal. Among the leading exporters of steam coal are Indonesia, Australia, Russia, Colombia, and South Africa. Indonesia, which was the world's largest exporter of steam coal in 2013, is expected to remain the top exporter through 2040. The three top exporters of coking coal are Australia, the US, and Canada. Despite a substantial drop in coking coal exports from the US, it is projected that Australia, the US, and Canada will continue to be the top three significant exporters of coking coal through 2040 (International Energy Outlook 2016, 2016).

The Coal Sector in the United States

The US coal industry has been and will continue to be during the coming years, the leading industry within the country's electricity generation sector. In the last quarter of the 20th century, the US economy grew, and the standard of living of the American people increased as a result of the considerable expansion of the national energy system. This expansion was carried out taking advantage of the vast coal resources in the US and its low price, which was cheaper than the cost of other energy sources available in the country at that time. The low price of coal allowed the leading coal companies to produce large quantities of this type of energy source, thus increasing the supply of coal to meet its growing demand in the electricity generation sector. "By many measures—productivity improvements,[4] coal-mining output, and growing share of power production—the US coal industry looked to be reasonably strong as of the year 2000" (Tierney, 2016).

In that year, over 90% of the US coal production was used for electricity generation and heating in the country. This situation changed after the

---

[3] In the case of North Korea, the application of the United Nations Security Council sanctions against the country would reduce significantly the exports of all types of coal to other countries.

[4] It is important to stress that this productivity improvements, however, did not come without consequences. Between 1985 and 2000, the number of US coal-mining jobs dropped by more than a half (US Department of Labor, Mine Safety and Health Administration, 2017). Over 2.100 mine workers lost their lives due to mining-related injuries between 1997 and 2000 alone (US Department of Labor- Bureau of Labor Statistics, 2018). Without a doubt, the use of coal as fuel in coal-fired power plants for electricity generation and heating contributed significantly to increase the level of carbon emissions and air contamination in the US.

year 2000 as a result of international pressure against the main polluting countries, one of which is precisely the US, to reduce the levels of environmental pollution due to the use of coal for electricity generation and heating.

The dominance position of the coal industry within the electricity generation sector in the US is due to the price advantage of this type of energy source in comparison with the price of other energy sources used for the same purpose. The shale gas revolution, without a doubt a critical success in the country after 2000, has changed the market conditions within the US energy market. As a result of this new condition, gas production grew by one third since 2008 with a significant impact on the gas price, which registered a decrease of 70%.

However, it is important to highlight that this continued economic advantage of coal in comparison with other types of energy sources depends upon continuing productivity improvements in the mining industry. Since 2000, several factors have reduced the participation of the coal industry in the electricity generation sector. The factors are:

• declining coal-mining productivity,
• shifts in world demand for coal as a result of the adoption of several measures by different countries to reduce environmental contamination and a negative impact on the population,
• shale-gas revolution, which minimizes the price advantage of coal,
• increase in efficiency in the use of electricity by consumers,
• foreseeable overall flat demand in the power sector during the coming years,
• recent cost reductions in the renewable energy technology for electricity generation and heating, and
• poor investments made by large coal companies.

Finally, it is important to note that after 1990, only a minimal number of coal-fired power plants was built in the US and, for this reason, the coal industry has an aging number of coal-fired power plants in operation with very inefficient electricity generation units and without modern pollution controls. This situation of the US coal industry is responsible for the country being one of the most significant pollutants in the world.

Coal Reserves in the United States

Coal is the most abundant fossil fuel in the US. By far, coal resources are more abundant than natural gas and oil resources together. In 2017, coal was the primary energy source used for electricity generation and heating within the country.

Coal is a combustible black or brownish sedimentary rock, which is composed mainly of carbon, hydrogen, and oxygen. Coal is formed from organic plant matter decayed, compressed and altered by geological processes over millions of years. Because of differences in the development of coal concerning pressure, temperature, and time, there are different types of coal (Bönisch, Nicole), which are:

- lignite,
- subbituminous,
- bituminous, and
- anthracite.

In the US, all of these types of coal can be found. There are six great areas where abundant coal reserves can be located in the US, which are:

- Appalachian region,
- Interior,
- Northern Great Plains,
- Rocky Mountains,
- Pacific Coast, and
- Gulf Coastal Plain.

Regarding the coal ranks, bituminous is mined mostly in the east and midwest states; subbituminous is mined only in the western countries; lignite in Gulf coast states and North Dakota; and anthracite in small quantities in Pennsylvania.

According to the EIA database 2017, the US has the largest coal reserves and resources in the world (demonstrated coal reserves base 476 billion short tons in 2017[5]) and has enough recoverable coal reserves to last over 300 years at current rate of consumption. This coal reserves are more significant than the nearest competitor, Russia, and over twice that of China. According to the EIA source, the estimated recoverable coal reserves in the US totaled 254.896 million short tons.[6]

The EIA statistics show that more than half (55%) of US coal reserves are located in the west, of which Montana and Wyoming together account for 43% of the total US coal reserves. About 70% of these reserves are in

[5] The demonstrated coal reserve base is composed of coal resources that have been identified to specified levels of accuracy, and that may support economic mining under current technologies. The demonstrated coal reserve base includes publicly available data on coal that has been mapped and verified to be technologically minable.

[6] Estimated recoverable coal reserves includes coal in the demonstrated coal reserve base considered recoverable after excluding coal estimated to be unavailable because of land use restrictions, and after applying assumed mining recovery rates.

Table 4.2 Recoverable Coal Reserves (million short tons).

Country	Recoverable Coal Reserves in 2015 at Producing Mines	Average Recovery Percentage in 2015 (%)	Recoverable Coal Reserves in 2016 at Producing Mines	Average Recovery Percentage in 2016 (%)
United States	18.327	78,62	16.956	79,26

Source: EIA database 2017.

the top five producing states, which are Wyoming, West Virginia, Pennsylvania, Illinois, and Kentucky. The US government owns about 88 billion short tons (one-third of the total) of US domestic coal reserves, followed by Great Northern Properties Limited Partnership (20 billion short tons) and Peabody Energy Corporation (6,3 billion short tons). Altogether, the top three reserve owners account for about 44% of US coal reserves. In Table 4.2, US recoverable coal reserves in 2015 and 2016 are shown.

According to Table 4.2, the recoverable coal reserves in the US during the period 2015—16 decreased by 7,5% falling from 18.327 million short tons in 2015 to 16.956 million short tons in 2016. The US was and remains the most significant world coal reserves holder with around 22% of the total. The recoverable coal reserves in the US represents more than 90% of the recoverable coal reserves in the North American region, which has the second world largest recoverable coal reserves after Europe (around 28%).

The evolution of the recoverable coal reserves at producing mines in the US during the period 2012—16 is shown in Fig. 4.6.

From Fig. 4.6, the following can be stated: the US coal recoverable reserves at producing mines decreased by 9,2% during the period 2012—16

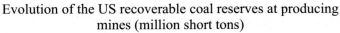

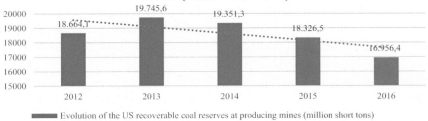

Evolution of the US recoverable coal reserves at producing mines (million short tons)

Figure 4.6 Evolution of the recoverable coal reserves at producing mines in the US during the period 2012—16 (million short tons). *(Source: EIA database 2017.)*

falling from 18.664,10 million short tons in 2012 to 16.956,4 million short tons in 2016. The peak in the US coal recoverable reserves at producing mines was registered in 2013. Since that year, the level of the US coal recoverable reserves at producing mines decreased each year continuously (total decreased 14,2%). It is predictable that this trend will continue during the coming years. The foreseeable decrease in the level of the recoverable coal reserves is due to a reduction of coal production in the US as a consequence of the measures adopted by the Trump administration on environmental issues, which are affecting the coal industry.

Coal Production in the United States

In the last decades, coal has been the primary energy source used for electricity generation and heating in the US. In 2017, coal-fired power plants generated 1.314 TWh, which represents 30,1% of the total electricity produced in the country in that year.

It also helped to provide many basic needs like transportation for people and heating. Coal production permanently grew since it was discovered and has a long tradition in the country. US coal production had been high since the 1990s and reached its highest level of production in 2008 (1.171,8 million short tons). After that year, coal production began to decline significantly. During the period 2015—17, the US coal production fell below 900 million short tons, and it is expected to continue this trend during the coming years. The evolution of the coal production in the US during the period 2012—17 is shown in Fig. 4.7.

As can be seen in Fig. 4.7, the production of coal in the US decreased by 23,8% during the period considered falling from 1.016,5 million short tons

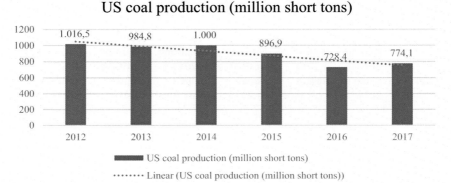

Figure 4.7 US coal production during the period 2012—17. *(Source: EIA database 2018.)*

in 2012 to 774,1 million short tons in 2017. During the period 2014—16, the coal production declined by 27,2% and reached the lowest annual level of production within the period 2012—17.[7] Five states were responsible for approximately 71% of the total US coal production in 2015, which are, according to the AGI American Geosciences Institute database (2017):

• Wyoming, with 41,9%,
• West Virginia, with 10,7%,
• Kentucky, with 6,8%,
• Illinois, with 6,3%, and
• Pennsylvania, with 5,6%.

Production in Western, Interior, and Appalachian areas decreased in the last years, and surface mines, from where the source of 66% of the total US coal production is extracted, accounted for 62% of the total number of mines in operation throughout the country. About 1,3 million short tons, or 0,2% of total coal production, was from coal waste recovery (AGI American Geosciences Institute database, 2017). It is expected that the participation of coal in the energy mix of the country will continue decreasing during the coming decades as a result of the new environmental measures adopted by the US administration.

As a result of international and national environmental regulations adopted to combat climatic changes, and despite the support that the new US administration is giving to the coal industry, the number of producing mines in the US continued its downward trend, falling from 985 mines in 2014 to 710 mines in 2016; this means a reduction of 28%.

The US coal production and the number of coal mines in operation throughout the country during the period 2014—16 is included in Table 4.3.

Table 4.3 US Coal Production and Coal Mines in Exploitation During the Period 2014—16.

Coal Production	Number of coal mines in 2014	Coal Production in 2014 (million short tons)	Number of Coal Mines in 2015	Coal Production in 2015 (million short tons)	Number of Coal Mines in 2016	Coal Production in 2016 (million short tons)
US total	985	1.000	853	896,9	710	774,1

Source: EIA database 2018.

[7] The level of coal production in the US, in 2017, is the lowest since 1978 (670,2 million short tons) (EIA database, 2018).

According to Table 4.3, the number of coal mines in operation in the US in 2016 decreased by 28% concerning the amount of coal mines that were operating in 2014; the coal production in the same period fell by 22,6%. It is expected that this trend will continue without change during the coming years. The reduction in the number of coal mines in operation in the US is, despite the support given to the coal industry sector by the new US administration:

- due to the implementation of the Clean Air Act of 1970,
- due to the struggle of the coal industry against the natural gas industry, and
- as a result of the increased use of renewable energy sources for electricity generation and heating, particularly solar and wind energy, and the reduction in the price of this type of energy sources during the last years.

Without a doubt, natural gas, solar, and wind energy are the preferred sources for electricity generation and heating in many countries, including the US, because they can produce electricity more cleanly and cheaply. However, the new energy policy adopted by the Trump administration to end the war against the use of coal for electricity generation and heating in the US could change this trend.[8]

According to EIA sources, the collective productive capacity[9] of US coal mines decreased by 6,3% during the last years. Capacity utilizations at underground mines and surface mines in 2015, both fell from the 2014 levels. The overall capacity utilization across all mines decreased from 80,1% to 76,7%; this means a reduction of 3,4%.

The US Department of Energy forecast that coal production will slip by 2% this year and next. That outlook is based on its projection that natural gas prices will remain low over much of the next two years and the demand for US coal exports will slow down during the same period. Different projections show that coal production would fall below 600 million short tons per year if the former Obama Administration's CPP is implemented. Without the implementation of the CCP, it is likely that coal production

---

[8] Coal production rose slightly during Trump's first year in office. The country produced 774,1 million short tons of coal in 2016. However, according to different sources, the main driver was steeper natural gas prices, which meant coal was more competitive than natural gas for electric power generation and heating.

[9] Collective productive capacity is the maximum amount of coal that can be produced annually, as reported by mining companies.

would remain relatively flat, between 800 and 900 million short tons per year during at least the next three decades.

It is important to highlight that 48 years after the adoption of the Clean Air Act of 1970, the closure of old and ineffective coal-fired power plants has finally begun. The coal-fired power plants retired in 2015 were quite old (more than 40 years of operation). For this reason, it can be said that the decline in coal-fired electricity generation in the US is mostly the result of the closure of ancient, ineffective, and contaminated coal-fired power plants still in operation in the country.

Despite the closure of several old and ineffective coal-fired power plants in the US during the last years, the participation of coal in the US energy matrix is expected to be 32% during the coming years, while the involvement of natural gas is likely to be only a little bit higher (33%) (Kolstad, 2017).

Coal Consumption in the United States

At the beginning of the 19th century, some of the first commercial coal mines in the world were already operating in various locations of the US. Throughout the 19th century the consumption of coal in the US grew steadily, and by the end of the century, the amount of energy produced by coal exceeded the amount of energy produced by wood. Industrialization, the use of coal for electrical machinery and for electricity generation and heating, had a significant impact on the growth of coal demand in the country.

During the 20th century, the use of coal for electricity generation and heating continued to grow in the US and coal became the primary energy source in the country, even though it is the most pollutant energy source used for electricity generation and heating with a significant negative impact on the environment and population. Coal is also one of the leading energy sources responsible for climatic changes. With the aim of reducing the adverse effects on the environment through the use of coal for electricity generation and heating in the US, the former Obama administration adopted a group of measures to reduce emissions by the energy sector nationwide to 30% below the 2005 level by 2030. These measures could have a significant effect on the coal-producing and coal-consuming industry. However, and despite the adoption of these measures by the US government, overall, coal is and will continue to have significant participation in the US energy mix at least during the coming decades (World Energy Resources, 2016 used with the permission of the World Energy Council, www.worldenergy.org, 2016).

Coal consumption in the US was consistently over one billion short tons per year from 2000 until 2011, except the year 2009. The peak in the US coal consumption was achieved in 2007 at 1.128 billion short tons. In 2016, the US coal consumption was 731.071 thousand short tons, a decline of 8,5% concerning 2015 (798.115 thousand short tons). Coal consumption in the US electric power sector in 2016 decreased by 59,9 million short tons reaching 678,6 million short tons. Coal consumption in the industrial, commercial, and institutional sectors was 52,5 million short tons lower in 2016 than in 2015 (a 12% lower), according to EIA sources.

The evolution of the coal consumption in the US during the period 2010—17 is shown in Fig. 4.8.

According to Fig. 4.8, the use of coal for electricity generation and heating in the US decreased by 31,7% during the period considered, falling from 1.048,51 thousand short tons in 2010 to 716,96 thousand short tons in 2017, as a result of the application of the former Obama energy policy on the reduction in the use of coal for environmental reasons. However, this trend would change during the coming years as a result of the implementation of the new energy policy adopted in 2017 for the Trump administration. The new energy policy eliminated a group of restrictions taken early by the former Obama administration on the use of coal for electricity generation and heating with the aim of reducing the emission of CO_2. Due to this change, the current role of coal in the energy mix of the country it is expected to be almost the same during the coming years (32%).

The EIA projects coal consumption to remain below 800 million short tons in the short–term and to drop close to 500 million short tons in the long–term if the new US administration implements the CPP.

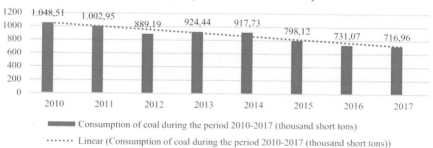

Consumption of coal during the period 2010-2017 (thousand short tons)

••••••• Linear (Consumption of coal during the period 2010-2017 (thousand short tons))

Figure 4.8 Evolution of the consumption of coal in the US during the period 2010—17. (*Source: EIA database 2018.*)

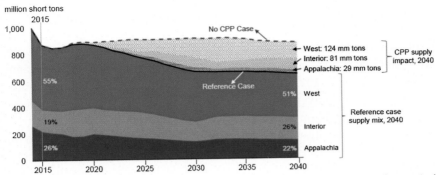

Figure 4.9 Evolution of the coal supply by region in the US during the period 2015–40. *(Source: Annual Energy Outlook 2016.)*

Without the implementation of the CPP, the US coal consumption level would rise slightly above 800 million short tons until 2030, and remain there through 2050.

According to Fig. 4.9, in 2015 the structure of the regional coal supply was the following: West 55%, Interior 19%, and Appalachia 26%. It is projected that in 2040 the structure of regional coal supply will be the following: West 51%, a reduction of 4% with respect to 2015; interior 26%, an increase of 7% with respect to 2015; and Appalachia 22%, a decrease of 4% with respect to 2015.

Measures Adopted to Reduce the Production and Consumption of Coal in the United States

Undoubtedly, the most drastic regulations adopted by the US government in recent years on the reduction of CO_2 emissions were those approved by the government of former President Obama, ordering a decrease of 30% by 2030 by those power plants that use fossil fuels for power generation, particularly coal. It has been anticipated that the application of these regulations will have a negative impact on the US economy, will cause the loss of almost a quarter of a million jobs per year, and the closure of the older and less efficient coal-fired power plants for electricity generation and heating throughout the country. The cost of the implementation of these measures is expected to be around US$50 billion per year.

These regulations will also impact consumers directly, as they will result in increased electricity prices, taking into account that the US depends on coal for 40% of its electricity. By adopting these regulations, the US, one of the largest polluting countries in the world, expects other countries, particularly China, to act similarly. The aim is a global reduction of current pollution levels that are very high and worrying. The implementation of these regulations is expected to lead to an estimation of between US$55 billion to US$93 billion per year in 2030 in climate and health benefits (Potenza, 2017).

The above-mentioned regulations, while having a negative impact on the coal industry and electricity generation and heating using this type of energy source as fuel, will support the development and use of other less polluting energy sources such as natural gas, nuclear energy, different types of renewable energy, particularly solar and wind energy, and will increase energy efficiency in electricity generation and heating. The critical points of the proposed regulations are:

- Carbon emissions are expected to be 30% below 2005 levels by 2030.[10] These are front-loaded reductions. Carbon emissions are expected to be 15% below 2005 levels, with a preferred target of a 25% reduction by 2020; a valid date in 2018, the industry will only have a few years to achieve a further 10% reduction.
- Individual states are mandated to cut emissions based on their energy mix. However, the regulations adopted encourages multistate programs to do so.
- States will be required to achieve interim targets between 2020 and 2029; the final compliance deadline is 2030.

According to government sources, carbon dioxide emissions constitute a national health crisis, in addition to impacting the economy and causing global warming.

In October 2017, the Trump administration announced that it would take steps to repeal a federal policy adopted by the former Obama administration that would:

- have pushed states to reduce the use of coal for electricity generation and heating significantly;
- lead to the closure of the oldest and less efficient coal-fired power plants used for electricity generation and heating; and
- switch to the use of different renewable energy sources for electricity generation and heating.

---

[10] President Trump had signed an executive order in March 2017 repealing the CPP, which the administration sees as an overreach in presidential power that kills jobs. In reality, the CPP was Obama's attempt at tackling climatic changes by ordering fossil fuel-fired power plants, which are the largest concentrated source of CO_2 emissions in the US, to cut carbon pollution by about 30% by 2030.

According to Trump administration, former Obama administration overstepped its legal authority by forcing utilities around the country to reduce carbon emissions outside their facilities by replacing coal-fired power plants with wind farms and solar parks, for instance. The proposal does not lay out a replacement to the CPP. Instead, the Trump administration wants to open a period of public debate on how to replace the CPP most effectively and, based on the outcome of these consultations, to cut emissions from fossil fuel-fired power plants, mainly coal-fired power plants (Potenza, 2017).

Coal Exports by the United States

One of the US coal industry's greatest challenge is to increase sales to overseas coal markets, particularly for steam coal, with the purpose of compensating the decline in domestic demand. The rise in US coal exports through 2012 was aided by a drop in the value of the US dollar against other currencies, including those of other major coal-exporting countries, such as Australia, Indonesia, and Russia.

The evolution of the coal exports from the US during the period 2000—17 is shown in Fig. 4.10.

According to Fig. 4.10, the US exports of coal during the period 2010—17 increased by 18,6% rising from 81.715,70 thousand short tons in 2010 to 96.953,4 thousand short tons in 2017. However, the US exports of coal decreased significantly (52,1%) during the period 2012—16, after a period of increase three years earlier. The peak in the US exports of coal

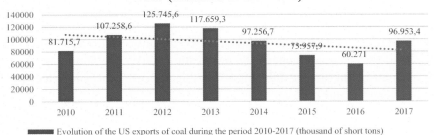

Figure 4.10 US exports of coal during the period 2010—17. *(Source: EIA database 2018.)*

during the whole period considered was achieved in 2012. It is predictable that the US exports of coal will continue to decrease during the coming five years despite the increase registered in 2017. For this reason, no significant additions to export mining capacity are expected to begin over the next five years in the country because of the projected weak domestic coal demand and low international prices.

Coal exports from the US to Asian markets are currently limited by scarce port capacity at the US west coast. To alleviate the problem due to lack of coal storage capacity in the US, west coast projects like the Gateway Pacific are underway with a planned export capacity between 24 and 38 million tons per year, including the Millennium Bulk Logistics project and the Port Westward project. Both projects have a projected capacity of between 15 and 30 million tons per year (Medium-Term Coal Market Report, 2015).

On the other hand, the decline of US coal exports from 97.256,7 thousand short tons in 2014 to 60.271 thousand short tons in 2016, a reduction of 38,1%, promoted the growth in Colombian coal exports, which hit a record of 83,3 million tons in 2016.

The ongoing limited US coal exports capacity and the replacement of old and inefficient coal-fired power plants with natural gas-fired power plants with the aim of reducing CO_2 emissions in the energy sector nationwide to 30% below the 2005 level by 2030, could have a considerable effect on the coal industry. However, and despite the negative impact on the coal industry as a result of the measures adopted by the US administration, it is likely that coal will continue to be a significant component of the US energy mix at least for the coming years, but its participation will be less than before.

Finally, it is important to highlight, that in the last years, more coal mines with high production costs were closed down, most of them in the US, Australia, and China. At the same time, all coal producers were focusing on cost-saving initiatives and improving their productivity in coal mining (World Energy Resources, 2016 Coal, used with the permission of the World Energy Council, www.worldenergy.org, 2016). Thus, it seems that the global oversupply situation in coal markets may hardly change shortly. Furthermore, reductions through coal mine closures are offset by the commissioning of new production capacities in some countries with substantial coal reserves.

Coal Imports by the United States

Although the US produces a large amount of coal (772 million short tons in 2017), some coal-fired power plants along the Gulf Coast and the Atlantic Ocean sometimes find it cheaper to import coal from other countries than to obtain this type of fossil fuel from US producing regions (Coal Explained. Coal Imports and Exports, 2018). Despite this fact, the US remained a net exporter of coal in 2017, exporting 96.953,4 thousand short tons and importing 7.777,2 thousand short tons.

The evolution of the imports of coal by the US during the period 2010—17 is shown in Fig. 4.11.

According to Fig. 4.11, the imports of coal by the US during the period 2010—17 decreased2 by 60%, falling from 19.352,7 thousand short tons in 2010 to only 7.777,2 thousand short tons in 2017. The peak in the imports of coal by the US during the period considered was reached in 2010. It is expected that the imports of coal by the US will continue to decrease at least during the next five years as a result of the closure of old and ineffective coal-fired power plants, and the reduction in the use of this type of energy source for electricity generation and heating.

It is important to highlight that the majority (90%) of coal imported into the US is steam coal, a type of coal that is primarily used for electricity generation and heating. Colombia remained the predominant source of US coal imports, despite a decrease of 12% (1 million short tons) in 2016. Metallurgical coal imports, primarily imported from Canada, fell by 44% (0,8 million short tons) in 2016.

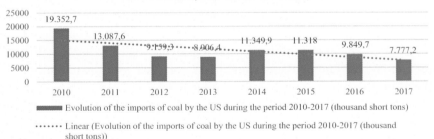

Figure 4.11 Evolution of the imports of coal by the US during the period 2010—17. *(Source: EIA database 2018.)*

Environmental Impact Due to the Use of Coal for Electricity Generation in the United States

Coal has long been a reliable source of the US energy sector, but it comes with tremendous costs because it is a high contaminating energy source. The same chemistry that enables coal to produce energy also produces some profoundly harmful pollutants that not only harm public health but also have a negative impact on the environment and population. Air pollutant and global warming are two of the most severe adverse effects on the environment and population as a result of the use of coal for electricity generation and heating. When coal burns, the chemical bonds holding its carbon atoms in place are fragmented, not only releasing energy but generating other chemical reactions, many of which carry toxic airborne pollutants and heavy metals into the environment (Coal and Air Pollution, 2017). According to the above-mentioned report, air pollutants include:

1. **Mercury**: Coal plants are responsible for 42% of US mercury emissions. According to the Environmental Protection Agency (EPA), "US coal power plants emitted 45.676 pounds of mercury in 2014".

2. **Sulfur dioxide (SO_2):** Sulfur dioxide is formed as a result of the contact between the sulfur contained in coal and oxygen. SO_2 combined with other molecules in the atmosphere form small, acidic particulates that can penetrate human lungs and it is "linked with asthma, bronchitis, smog, and acid rain. SO_2 damages crops and other ecosystems, and acidifies lakes and streams. US coal power plants emitted more than 3.4 million tons of SO_2 in 2014" (Coal Air pollution, 2017).

3. **Nitrogen oxides (NOx):** Nitrous oxides are visible as smog. Exposure of humans to NO_x results in "irritation of lung tissue, exacerbating asthma, and making people more susceptible to chronic respiratory diseases like pneumonia and influenza." (Coal Air Pollution, 2017). In 2014, US coal power plants emitted more than two million tons of NO_x into the atmosphere;

Other harmful pollutants emitted in 2014 by the US coal-fired power plants include 41,2 tons of lead, 9.332 pounds of cadmium and other toxic heavy metals, 576.185 tons of carbon monoxide, 22.124 tons of volatile organic compounds, 77.108 pounds of arsenic, and 197.286 tons of small airborne particles (Coal Air Pollution, 2017).

According to EIA sources (Coal and Environment 2018), in 2016, "surface mines (sometimes called strip mines) were the source of about 65% of the coal mined in the United States. These mining operations remove

Table 4.4 Evolution of the Emission of Contaminated Gases to the Atmosphere in the US (thousand metric tons).

Year	Carbon Dioxide (CO_2)	Sulfur Dioxide (SO_2)	Nitrogen Oxides (NOx)
2010	2.388.596	5.400	2.491
2011	2.287.071	4.845	2.406
2012	2.156.875	3.704	2.148
2013	2.173.806	3.609	2.163
2014	2.168.284	3.454	2.100
2015	2.031.452	2.548	1.824

Source: EIA database 2016.

the soil and rock above coal deposits or seams". Mountaintop removal and valley fill is a technique with which "the tops of mountains are removed using explosives. This technique changes the landscape, and streams are sometimes covered with rock and dirt". The water pump out from these filled valleys may contain different types of contaminants that can harm aquatic wildlife downstream. "Underground mines generally have a lesser effect on the landscape compared to surfaces mines. However, the ground above none tunnels can collapse, and acidic water can drain from abandoned underground mines" (Coal and Environment, 2018).

The evolution of the emissions of different contamination gases in the US during the period 2010—15 is included in Table 4.4.

According to Table 4.4, during the period 2010—15, there has been a reduction in the emission of the three types of the main contaminated gases to the atmosphere in the US. In the case of CO_2, the decrease was 25%; of SO_2, 53%; and of NOx 27%.

The Impact of Energy Regulations and the Global Coal Situation of the Coal Industry in the United States

The evolution of coal power in the US in the last three years accounted for 39% of the country's electricity production at utility-scale facilities in 2014; 33% in 2015; and 30,4% in 2016 (EIA database, 2017).

Since 2012, a total of 50 US coal companies have filed for bankruptcy protection, underlining how complicated and difficult the situation has become for the US coal industry, and also for the global coal industry as well, due to the environmental regulations adopted by the US administration and at the world level as well. In 2016, Arch Coal, the US's fourth-biggest coal company, declared bankruptcy in August; Walter Energy announced insolvency in July; and Patriot Coal made its second filing since 2012 for

Chapter 11 protection in October, before being liquidated and sold off to other coal producers (Springfield, 2016).

Global coal prices have fallen consistently during the period 2014–20, thanks mainly to the declining demand for coal, global coal oversupply, and a spate of structural and environmental factors dragging the industry down. Given that China is the world's leading coal consumer, the decision of the Chinese government to close the country's most toxic coal-fired power plants in the coming years, and the adoption of a moratorium in the opening of new coal-fired power plants until at least 2019 is not good news for the US coal industry. There is no way to compete in the Asian market given the transportation costs US exporters have to bear, which other countries in closer proximity, such as Indonesia, do not require. Back in 2010, global coal prices were high enough for US exporters to make a profit on the Asian market. Today, this is merely unachievable (Springfield, 2016).

Summing up the following can be stated: in the US, the coal industry is declining as a result of the Environmental Protection Agency policies and low natural gas prices. These policies will require existing power plants to cut carbon emissions by 30% in 2030. Since 2010, utilities have formally announced the retirement of a significant number of old and inefficient coal-fired power plants, and it is expected also to see more power plants of this type closed or substituted with natural gas-fired power plants by 2020. However, the new US administration policy on energy matter could stop this decline and would try to keep the participation of coal in the US energy mix in the future as high as possible.

Environmental Impact and Technological Response in the United States

Global energy consumption raises some environmental considerations, particularly in the case of use of fossil fuel for electricity generation and heating. In the specific case of coal, the release of different pollutants, such as SOx, NOx, and mercury, among others, has a substantial negative impact on the environment and population. With the purpose of reducing the emission of these and other pollutants, new technologies associated to the use of coal for electricity generation and heating have been developed and deployed with the aim of minimizing the emissions of the above-mentioned contaminants.

The release of CO_2 into the atmosphere from human activities has been linked to global warming and climatic changes. The combustion of fossil fuels for electricity generation and heating is an important source of

CO_2 emissions worldwide. While the use of oil in the transportation sector is the primary source of energy-related CO_2 emissions, the use of coal for electricity generation and in the industry is also a significant source of this type of pollutant gas emission. As a result, the industry has been researching and developing technological alternatives to reduce the current level of CO_2 emissions (World Energy Resources, 2016 Coal, used with the permission of the World Energy Council, www.worldenergy.org, 2016).

Clean coal technologies are a range of technological options, which improve the environmental performance of coal used for electricity generation and in the industry. These technologies reduce emissions and waste and increase the amount of energy gained from each ton of coal consumed. Different techniques suit different types of coal and tackle different environmental problems. The choice of which is the best technology to be used depends on a country's level of economic development.

A major environmental challenge facing the world today is the risk of global warming and climatic changes. The IEA advocates a two-step process of reducing emissions from coal: firstly, by improving power plant thermal efficiency while providing meaningful reductions in CO_2 emissions, and secondly by advancing new technologies to the commercial scale. Improving the efficiency level increases the amount of energy that can be extracted from a single unit of coal. Increases in the efficiency of electricity generation and heating are essential in tackling climate changes. One percentage point improvement in the effectiveness of a conventional pulverized coal combustion power plant results between 2 and 3% reduction in CO_2 emissions. It is important to highlight that highly efficient modern supercritical and ultrasupercritical coal-fired power plants emit almost 40% less CO_2 than subcritical plants (World Energy resources 2016 Coal, used with the permission of the World Energy Council, www.worldenergy.org, 2016).

According to the Carbon Capture and Storage Association, carbon capture and storage is the process of capturing (90%) of CO_2 from large point sources, transporting it to a storage site, and depositing it where it will not enter the atmosphere, usually an underground geological formation. There are three methods of carbon capturing: (1) precombustion capture, (2) postcombustion capture, and (3) oxyfuel combustion.

Efficiency improvements include the most cost-effective and shortest lead time actions for reducing the emissions of CO_2 and other contaminating gases produced by the operation of coal-fired power plants used for electricity generation and heating. This is particularly important for developing countries and economies in transition where existing power plant efficiencies are generally lower than the ones operating in advanced countries. The average global effectiveness of coal-fired power plants is currently 28% compared to 45% for the most efficient power plants. A program of repowering existing coal-fired power plants to improve their efficiency, coupled with the newer and more efficient coal-fired power plants being built, will generate significant CO_2 reductions of around 1,8 Gt annually. Although the deployment of new, highly efficient coal-fired power plants is subject to local constraints, such as ambient environmental conditions and coal quality, deploying the most efficient coal-fired power plants possible is critical to enable these plants to be retrofitted with the carbon capture technology in the future (World Energy Resources 2016 Coal, used with the permission of the World Energy Council, www. worldenergy.org, 2016).

Improving the efficiency of the oldest and most inefficient coal-fired power plants would reduce CO_2 emissions by almost 25%, representing a 6% reduction in global CO_2 emissions since the adoption of the Kyoto Protocol. By way of comparison, under the Kyoto Protocol, parties have committed to reducing their CO_2 emissions by at least 5%. These emission reductions can be achieved by the closure of coal-fired power plants with a generating capacity higher than 300 MW and with more than 25 years of operation, and its replacement with larger and markedly more efficient coal-fired power plants. Where technically and economically appropriate, the replacement or repowering of larger old and inefficient power plants with high-efficiency plants of more than 40% should be encouraged. It is also important to note that the cost of reducing emissions from more efficient coal-fired power plants used for electricity generation and heating can be very low, requiring relatively small additional investments. This is especially the case when compared to the cost of avoided emissions through the use of different renewable energy sources and nuclear energy (World Energy Council 2016 Coal, used with the permission of the World Energy Council, www.worldenergy.org, 2016).

Electricity Generation in the United States Using Coal as Fuel

Coal has been used to generate electricity in the US since an Edison coal-fired power plant was built in New York City in 1882. The first alternating current power plant was opened by General Electric in Ehrenfeld, Pennsylvania in 1902, servicing the Webster Coal and Coke Company (Speight, 2012). By the middle of the 20th century, coal became the leading fossil fuel used for electricity generation and heating in the US. The long, steady rise of coal-fired electricity generation shifted to a decline after 2007. The decline has been linked to the increased availability of natural gas, decreased consumption (Houser et al., 2017), increased use of renewable power, and more stringent environmental regulations. The US Environmental Protection Administration has advanced restrictions on coal-fired power plants to counteract mercury pollution, smog, and global warming.[11]

The evolution of the net electricity generation capacity in the US using coal as fuel during the period 2010—17 is shown in Figs. 4.12 and 4.13.

According to Fig. 4.13, the net electricity generation in the US using coal as fuel during the period 2010—17 decreased by 35%, falling from 1.847.290 thousand MWh in 2010 to 1.207.901 thousand MWh in 2017. In that year, 30,1% of the total electricity generated in the country (around 4.015 billion kWh) was produced using coal as fuel. It is expected that this trend will continue without change in the future. However, as a result of

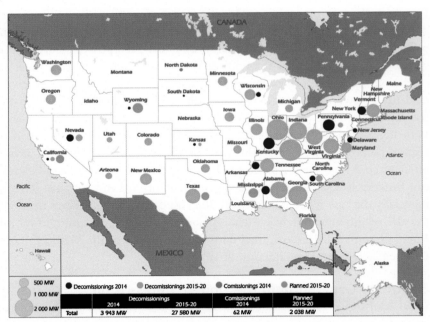

Figure 4.12 Coal-fired power plant projects and decommissioning in the US during the period 2014—20. (*Source: World Energy Outlook 2015.*)

[11] According to the US EIA, 27 GW of capacity from coal-fired generators have been retired from 175 coal-fired power plants between 2012 and 2016 (Gerhardt, 2012).

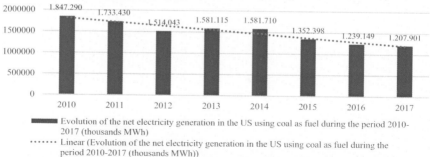

Figure 4.13 Evolution of the net electricity generation in the US using coal as fuel during the period 2010—17 (thousands MWh). *(Source: EIA database 2018.)*

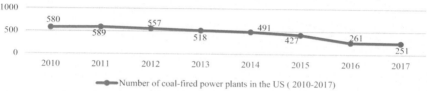

Figure 4.14 Number of coal-fired power plants in operation in the US during the period 2010—17. *(Source: EIA database 2018 and the Electric Power Monthly 2018 report.)*

the new energy policy adopted, in 2017, by the Trump administration, perhaps this trend would slow down during the coming years.

The number of coal-fired power plants in operation in the US during the period 2010—17 is shown in Fig. 4.14.

According to Fig. 4.14, the number of coal-fired power plants used for electricity generation and heating in the US decreased by 56,8% during the period considered, falling from 580 coal-fired power plants in operation in 2010 to 251 coal-fired power plants in service in 2017 (Fig. 4.15). It is expected that this trend will continue at least during the coming years but at a slower pace (Fig. 4.16).

From the socio-economic point of view, coal has been a support for the economy in both developed and developing countries alike for many years. Despite all efforts made in several countries to provide electricity to their population, there are still over 1,2 billion people in the world who live without adequate power, which is vital for basic needs. Electrification is a

Figure 4.15 View of coal Widows Creek power plant. *(Source: Tennessee Valley Authority Wikimedia Commons.)*

critical component in the economic and social development of all countries, independently of the region where they are located. The ability to provide reliable electricity has far-reaching effects on the economic and social growth in any country. Electrification leads to advancements in public health, education, transportation, communications, manufacturing, and trade (World Energy Council 2016 Coal, used with the permission of the World Energy Council, www.worldenergy.org, 2016). In some countries, access to electricity is a fundamental social right, and yet the demand for electricity continues to outstrip some regions' ability to supply it because of a lack of fuels, transmission, or infrastructure (The Socioeconomic Impacts of Advanced Technology Coal-Fueled Power Stations, 2015).

In many cases, achieving electrification of some remote areas in a country would not be possible without the use of coal as fuel. Its role in the electricity system in several countries is an essential factor in ending electricity poverty for billions of people and contributing to their economic and social development, despite its negative impact on the environment and population. Coal can also play a significant role in global steel production.

According to recent statistics issued by the World Steel Association (2015), there has been an increase in global steel production during the period 2010—14, reaching 1.665 million tons in the last year on the mentioned period, which represented a 16,2% increase from 2010 values. Coking coal is an essential element in blast furnace steel production, making up 70% of total steel production (the remainder is produced from electric arc furnaces using scrap steel).

There are socio-economic benefits and also concerns with regards to the use of coal for electricity generation and heating. Firstly, one can look at the benefit of coal mining in rural and remote areas where transport infrastructural development becomes the norm, since roads or rail lines need to be present for the transfer of coal to the power plants that use it for electricity generation and heating. The impact of coal on infrastructure development is more noticeable in developing nations than in developed countries, due to the absence of preexisting infrastructure. A variety of industries can also utilize the rail lines used to transport coal for the transport of materials for manufacturing their products. The investment in infrastructure caused by the energy industry helps to foster the economic and social development of the region involved in this investment. Furthermore, the local population will benefit since employment is provided and hence, other businesses will begin to prosper owing to the increase in market transactions and needs (World Energy Council 2016 Coal, used with the permission of the World Energy Council, www.worldenergy.org, 2016).

On the other hand, concerns can also be seen in that the natural topography of land close to the mining area is disrupted and disfigured. Furthermore, air quality significantly deteriorates as coal dust particles linger in the atmosphere as a result of mining activities and during the electricity generation and heating; however, this is mainly due to poor emission control. Another effect of poor management practice is the change that mining brings to groundwater, as the watercourse is diverted for the extraction process to occur (Mutemi, 2013). This often has an adverse effect on communities that depend on the use of underground water to sustain their source of income or for survival.

The Capacity Factor

Capacity factor measures the overall utilization of a power-generation facility or fleet of generators. Capacity factor is the annual generation of a power plant (or fleet of generators) divided by the product of the capacity and the number of hours over a given period. In other words, it measures a power plant's actual generation compared to the maximum amount it could generate in a given period without any interruption. As power plants sometimes operate at less than full output, the annual capacity factor is a measure of both how many hours in the year the power plant operated and at what percentage of its entire production.

Table 4.5 Evolution of the Capacity Factors for Conventional Power Plants During the Period 2013–17.

Year	Coal (%)	Natural Gas-Fired Combined Cycle (%)	Natural Gas-Fired Combustion Turbine (%)	Natural Gas Steam Turbine (%)	Natural Gas Internal Combustion Engine (%)	Petroleum Steam Turbine (%)	Petroleum Liquids-Fired Combustion Turbine (%)	Petroleum Combustion Engine (%)
2013	59,8	48,2	4,9	10,6	6,1	12,1	0,8	2,2
2014	61,1	48,3	5,2	10,4	8,5	12,5	1,1	1,4
2015	54,7	55,9	6,9	11,5	8,9	13,3	1,1	2,2
2016	53,3	55,5	8,3	12,4	9,6	11,5	1,1	2,6
2017	53,5	54,8	9,4	11,3	N.A.	13	2,0	N.A.

Source: EIA database 2018.

The annual capacity factor of a power plant is, therefore, a measure of availability (how much hours it is available to generate electricity) and an indirect measure of the marginal cost of generation (for non–variable sources) and other characteristics such as flexibility and startup times.

The evolution of the capacity factors for conventional power plants during the period 2013–17 is shown in Table 4.5.

According to Table 4.5, the capacity factor of coal-fired power plants takes second place with regards to the capacity factor of the natural gas-fired combined cycle since 2015. These two types of power plants have the highest capacity factor concerning other fossil fuel power plants within the period 2013–17. At the same time, the capacity factor of coal-fired power plants has been decreasing from 54,7% in 2015 to 53,5% in 2017. It is expected that this trend will continue without change at least during the coming years.

The Coal Sector in Canada

Canada is a midsized coal-producing country, ranked 12th at the world level in 2016. The annual production is a 50%-50% split between coking and thermal coal. Canada is the world's third-largest seaborne coking supplier and is a net exporter of energy, after Australia, and the US. The indigenous production of oil, natural gas, and coal currently exceed the consumption within the country. For this reason, exports of oil, natural gas, and coal play a vital role in the Canadian economy as a whole, and it is expected to play the same role during the coming years.

However, in the specific case of coal, it is expected that this type of fossil fuel will play a relatively minor role in Canada's energy supply system in the future, due to the adoption by the government of strong measures to protect the environment and population. For this reason, the role of coal within the energy supply system is expected to decline further with federal and provincial government efforts to reduce emissions of CO_2 and several other contaminating gases.

Canada's total coal consumption is expected to decline by 51% (0,4 quadrillions Btu) during the period 2012–40, and the share of coal in total primary energy supply is likely to decrease from 5% in 2012 to 2% in 2040; this means a reduction of 3%. In 2012, more than three-quarters of the coal consumed in Canada was used to generate electricity, with most of the rest going to industrial plants. The elimination of coal-fired generation in the Ontario province in April 2014, followed by enactment of the Canadian government entitled "Reduction of Carbon Dioxide Emissions from Coal-fired Generation of Electricity" regulations on July 1, 2012, is likely to result in more closures of coal-fired power plants within the country during the coming years (Creating Cleaner Air in Ontario, 2014; Reduction of Carbon Dioxide Emissions from Coal-fired Generation of Electricity Regulations, 2012). Consequently, in the International Energy Outlook 2016 with Projections to 2040 report, the electric power sector share of Canada's total coal consumption is projected to fall to 36% in 2040, and the coal share of the whole electricity generation and heating is expected to decline from 10% in 2012 to 1% in 2040; this means a significant reduction of 9%.

Coal Reserves in Canada

According to the Canada Energy Policy Laws and Regulations Handbook (Volume 1, 2015), Canada has the 10th largest coal reserves in the world, an enormous amount considering the sparse population of the country. However, the vast majority of those coal reserves are located very far away from the country's industrial centers and seaports. The exploitation of these largest coal reserves will increase transportation costs considerably and, therefore, will be uneconomical. For this reason, most of these coal reserves remain mostly unexploited. As happens with other natural resources, regulation of coal production is within the exclusive authority of the provincial governments. Federal regulations are applied only in the case when coal is imported or exported from Canada. Over 90% of Canada's coal reserves and 99% of its production are carried out in the three western provinces of Alberta, British Columbia, and Saskatchewan.

Of the above three provinces, Alberta alone has 70% of Canada coal reserves, and coal deposits underlay 48% of the Texas–sized territory. British Columbia has one of the thickest coal deposits in the world, the Hat Creek deposit, which is 550 m (1.800 feet) thick. There are also smaller, but substantial, coal deposits in the Yukon, the Northwest Territories, and the Arctic Islands. The Atlantic provinces of Nova Scotia and New Brunswick have coal deposits that were historically an essential source of energy, and Nova Scotia was once the largest coal producer in Canada. However, these coal deposits are much smaller and much more expensive to produce than the coal located in the west of Canada and, for this reason, coal production in the Atlantic province has virtually ceased (Energy Policy of Canada, 2018).

Number of coal mines in Canada (2017)

Figure 4.16 Number of coal mines in Canada. *(Source: Statistics Canada (2018).)*

Canada had 24 large coal mines, 19 of which are currently operating (79,2% of the total). There were 10 coal mines in British Columbia, 9 in Alberta, 3 in Saskatchewan, and 2 in Nova Scotia (see Fig. 4.16). There are no government-owned or operated coal mines in Canada; all are owned and operated by the private sector.

Also, with the number of coal mines operating in the country, there are many more projects in the exploration phase or advanced stages of regulatory approval. A total of 91,6% of Canada's coal deposits are located in western provinces (22 in British Columbia, Alberta, and Saskatchewan). Canada coal proved reserves are shown in Table 4.6.

By the data included in Table 4.6, the following can be stated: in 2016, hard coal proved reserves in Canada represented 1,9% of the North America region hard coal proved reserves and 6,8% of the lignite reserves; in 2016, total coal proved reserves in Canada represented 2,5% of the total coal proved reserves in the North America region.

The evolution of coal proved reserves in Canada during the period 2010—17 is shown in Fig. 4.17.

According to Fig. 4.17, the total coal proved reserves in Canada had been the same during the whole period considered. The country occupies the 17th place at the world level regarding the amount of proved coal reserves reported in 2016. The proved coal reserves of Canada represented, in 2016, around of 0,6% of the world proved coal reserves. More than 50% of the Canadian coal reserves are anthracite and bituminous coal; the remaining coal reserves are subbituminous and lignite coal.

Table 4.6 Coal Proved Reserves in Canada in 2017 (million tons).

Country/Region	Hard Coal	Lignite	Total
Canada	4.346	2.236	6.582
North America	228.330	32.770	261.100

Source: BP Statistical Review of World Energy 2016 and Statista database 2018.

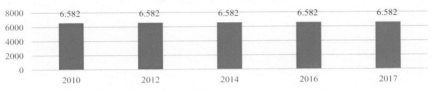

Evolution of coal proved reserves in Canada 2010-2017
(billion metrics tons)

Figure 4.17 Evolution of coal proved reserves in Canada during the period 2010–17. *(Source: Statista database 2018.)*

Coal Production in Canada

Canadian coal production declined or was flat during the period 2006–16. In the last 10 years, coal production in Canada reached a peak in 2013 of 36,4 million tons of oil equivalent. In 2016, coal production reached 31,4 million tons of oil equivalent, which is the lowest coal production registered by Canada during the period considered. It is projected that this trend will continue up to 2030, when old and inefficient coal-fired power plants should be closed, following a federal government decision adopted on this issue in 2016. The majority of Canadian coal production is in western Canada, namely British Columbia, Alberta, and Saskatchewan, according to the Canadian National Energy Board.

There are two types of coal produced in Canada: (1) thermal coal and (2) metallurgical coal Thermal coal is used primarily in electric power generation, particularly in Alberta and Saskatchewan provinces, reaching 93% of the total coal consumption in 2015.[12] Metallurgical coal is primarily used for steel production. Currently, about half of the coal produced in Canada is thermal coal, and domestic power producers use most of it; the remainder is exported. The majority of Canadian metallurgical coal is shipped to Asia; the rest is used by domestic iron and steel-markets, according to the Canadian National Energy Board.

The evolution of coal production[13] in Canada during the period 2010–17 is shown in Fig. 4.18.

---

[12] According to Natural Resources Canada, the consumption of coal by coal-fired power plants generated around 18% of the total electricity produced in the country in 2014. It is expected that this participation will decrease during the coming years.

[13] Fig. 4.18 includes commercial solid fuels, i.e., bituminous coal and anthracite (hard coal), lignite and brown (subbituminous) coal, as well as other commercial solid fuels.

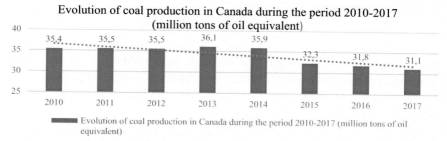

Figure 4.18 Evolution of the Canadian coal production during the period 2010—17 (million tons of oil equivalent). *(Source: BP Statistical Review of World Energy 2018 (June 2018).)*

From Fig. 4.18, the following can be stated: during the period considered, coal production in Canada decreased by 12,1% falling from 35,4 million tons of oil equivalent in 2010 to 31,1 million tons of oil equivalent in 2017. The peak in the production of coal within the period 2010—17 was registered in 2013. After that year, the output of coal decreased by 13,9%, which is around 2,5% lower than the production level recorded in the previous period. It is probable that as a result of the measures already adopted by the Canadian government, the production of coal in the country will continue to decrease until 2030 when all old and ineffective coal-fired power plants are projected to be closed.

The projected decline in coal production in Canada is attributed mainly to coal-fired power plant phase-out in Alberta. It is expected that nearly all the coal produced in Canada by 2040 will be metallurgical coal (32 million tons of oil equivalent or 94,1% of the total). It is also foreseeable that the demand for metallurgical coal used in the steel manufacturing sector will be stable until 2040.

Finally, it is essential to highlight the following: global coal markets and price trends will influence the production of coal in Canada during the coming years and the level of coal exports. If traditional coal importing countries move away from the use of coal for electricity generation and heating, then it is expected that coal demand and prices at the world level could decrease. However, it is also possible that the global coal demand for steel and electricity generation and heating could be higher than anticipated and, for this reason, this situation could lead to an increase in coal exports as well.

Coal Consumption in Canada

According to the International Energy Agency's World Energy Outlook 2016 with Projection to 2040 report, it is predictable that the overall coal demand will grow at 0,4% per year through 2040 because of the increased demand for thermal coal used in power plant. In the specific case of Canada, the coal demand has been decreasing systematically in last years and it is probably that this trend will continue without change during the coming years.

The evolution of the consumption of coal in Canada during the period 2010−16 is shown in Fig. 4.19.

According to Fig. 4.19, the consumption of coal in Canada during the period 2010−17 decreased by 25% falling from 24,8 million tons of oil equivalent in 2010 to 18,60 million tons of oil equivalent in 2017. By the decision adopted by the government of Canada to close all old and ineffective coal-fired power plants by 2030, it is anticipated that the consumption of coal for electricity generation and heating in the country will continue to decrease during the coming years as well as coal's participation in the country's energy mix.

It is important to stress that coal is mainly used in electricity generation and heating and in the industrial sector in Canada. Coal-fired power plants generated around 9,5% of the total electricity produced by the country in 2014. Electricity generation accounted for 79,5% of coal consumption in 2013 and 14,9% in the case of the industry. Coke ovens, other energy industries, and energy own-use accounted for 5,5% of coal consumption in 2013. The residential sector used 0,1% of the coal produced, a share which has been declining for decades (down from 2,6% in 1973).

Evolution in the consumption of coal in Canada (million tons of oil equivalent)

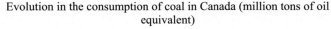

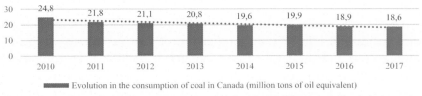

Evolution in the consumption of coal in Canada (million tons of oil equivalent)

Linear (Evolution in the consumption of coal in Canada (million tons of oil equivalent))

Figure 4.19 Evolution of the consumption of coal in Canada during the period 2010−16 (million tons of oil equivalent). Note: The data include solid commercial fuels only, i.e., bituminous coal and anthracite (hard coal), lignite and brown (subbituminous) coal, and other commercial solid fuels, and exclude coal converted to liquid or gaseous fuel, but include coal consumed in transformation processes. *(Source: BP Statistical Review of World Energy 2018 (June 2018).)*

The Canadian government projects that the power generation sector's share of total coal consumption will continue to decline during the coming decades. Declining coal consumption in power generation is unavoidable as a result of the provincial phase-out programs and federal emission regulations that have been already adopted.

In 2014, Ontario became the first province in Canada and in the world to close all coal-fired power plants. The coal installed capacity phase out represents a total of 7,5 GW or 35% of Canada's total coal-fired power generation.

Federal greenhouse gas regulations for coal-fired electricity generation will require each coal-fired unit that reaches a defined period of operating life (generally 50 years) to meet a performance standard that will expect them to be shut down, retrofitted with CCS, or converted to use a different fuel for electricity generation and heating. By 2030, more than half of Canada's coal-fired generating units will be subject to compliance with these regulations.

Coal Exports From Canada

According to Energy Policies of IEA Country Review (2015), Canada is a net exporter of coal and is the third-largest seaborne coking coal exporter in the world after Australia and the US. Almost half of Canada's coal production includes metallurgical (coking) coal, most of which is destined for export markets; the other half (mostly brown coal) was used in domestic coal-fired power plants for electricity generation and heating. Exports of coking coal registered, in 2016, an increase of 0,5% with respect to 2015 (in value the increase was of 30,7%). It is important to single out that exports of coking coal had a steady increase since 2009. The peak in the Canadian coal exports in the past 10 years was reached in 2013 when the country exported 39,1 million tons. The previous level of coal exports was driven by a high global demand for this type of fossil fuel. However, in 2014, due to a decreased need for coal, particularly in China, the exports of this type of fossil fuel in Canada dropped to 34,5 million tons. In 2016, coal exports further decreased to 30 million tons, a reduction of 13,1% with respect to the level reported in 2014. In 2016, Canada exported 70% of the total coking coal produced to Asian markets. The countries were Japan (24%), South Korea (20%), China (16%), and India (10%),[14] and these exports are

---

[14] Canada also exports coking coal to the US, Chile, and Brazil.

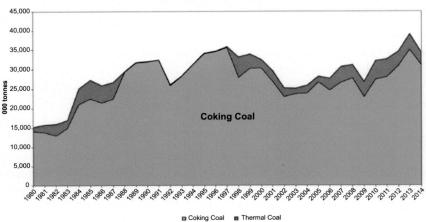

Figure 4.20 Development of coal exports in Canada. *(Source: Statistics Canada (2015).)*

rising due to high demand for this type of coal for steel production in these countries. Exports of thermal coal decreased by 5,9% by volume in 2016, while the value decreased by 10,9%. In 2017, Canadian coal exports reached US$5,1 billion or 4,6% of the world total coal exports.

Canadian coal exports are expected to increase at an annual growth rate of nearly 1,5% over the coming years, and it is supposed to reach about 34 million tons in 2020 (Fig. 4.20).

While the international seaborne coal trade grew over the last years, the global oversupply of coal (both brown and hard coal) led to a significant worldwide coal price reduction that affected all key coal exporters.

The evolution of coal exports by Canada during the period 2005—16 is shown in Fig. 4.21.

According to Fig. 4.21, the exports of coal from Canada during the whole period considered increased by 7,1% rising from 28 million tons in 2005 to 30 million tons in 2016. However, the main increase was reached during the period 2005—13 (42,9%) rising from 28 million tons in 2005 to 40 million tons in 2013. After that year the exports of coal by Canada dropped 25% falling from 40 million tons in 2013 to 30 million tons in 2016. It is predictable that the exports of coal from Canada will continue to decrease during the coming years, due to a reduction in coal demand at the world level.

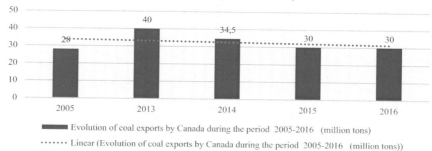

Figure 4.21 Evolution of coal exports by Canada during the period 2005–16. *(Source: Coal Facts, Natural Resources Canada (2018).)*

Coal Imports by Canada

Canada imports hard coal, both coking and thermal coal, for use in steel mills and coal-fired power generation. Most of the coking coal came from the Appalachian region, where it is geographically closer than the Canadian Rocky Mountains.

In 2016, half of the coal imports by Canada (metallurgical coal) was destinated for its use in steel manufacturing and the other half was thermal coal used in coal-fired power plants for electricity generation and heating. Imports of hard coal originated from the US represented 75,3% of the total coal imported by Canada in 2016, followed by Colombia with 21,5%, Ukraine with 2,6%, and Venezuela with 0,5%. Canada also imports brown coal, but imports of this type of coal have declined over the last decade. Imports of metallurgical coal and coke products also experienced a decrease in volume ($-13,7\%$) and value ($-24,5\%$). Similarly, imports of thermal coal saw a decline in size ($-15,4\%$) and value ($-10,5\%$). The decrease in coal imports by Canada in recent years is directly linked to the lower need for coal for power generation, according to the Energy Policies of IEA Country Review (2015).

On the other hand, it is important to highlight that in case that traditional coal importing countries reduce their demand of this type of energy source, particularly for electricity generation and heating during the coming years, it is also possible that world coal demand and prices associated to coal could decrease significantly. In this scenario, Canada could be forced to reduce coal exports and imports.

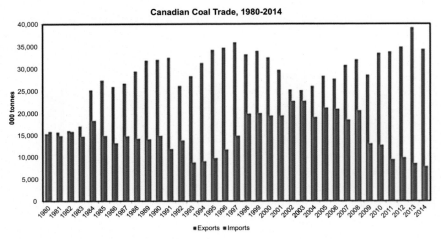

Figure 4.22 Development of trade (imports vs. exports) through 2014. *(Source: Natural Resources Canada, Statistics Canada (2015).)*

From Fig. 4.22, the following can be stated: after 1982, the level of coal imports by Canada was a little bit lower than the exports of this type of energy source. After that year, the level of coal exports was significantly higher than the imports of coal. After 2008, the exports of coal were considerably higher than the imports of this type of energy source, and the differences between these two indicators are increasing each year since 2009.

The evolution of coal imports by Canada during the period 2005—17 is shown in Fig. 4.23.

According to Fig. 4.23, coal imports by Canada during the period 2005—17 decreased significantly (68%) falling from 23,2 million tons in 2005 to 7,5 million tons in 2017. This reduction is the result of the adoption of a group of measures by the major coal importer, with the aim

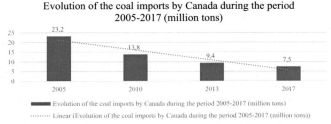

Figure 4.23 Evolution of coal imports by Canada during the period 2005—17. *(Source: Indexmundi and Coal Facts Natural Resources Canada (2017).)*

of reducing the use of coal for electricity generation and heating. It is expected, as a result of the Paris Convention on climatic changes, that this trend will continue in the future but perhaps at a slow rate.

Environmental Impact of the Use of Coal for Electricity Generation in Canada

Coal-fired power plants have long been an indispensable source of electricity in Canada, particularly in provinces without access to significant sources of hydroelectricity. But, while it has historically been considered as a low-priced option for electricity generation and heating, the use of coal for this specific purpose has many hidden costs and severe negative impacts on the environment and population. The adverse effects of the use of coal for electricity generation and heating are higher than the effect caused by any other energy source used for this specific purpose (Israël and Flanagan, 2016).

Coal-fired electricity generation produced a significant amount of greenhouse gases and is also a leading emitter of several air contaminants that are harmful to human health. Coal-fired power plants are a particularly important source of contaminated matters. The use of this antiquated technology will continue to produce GHG emissions that contribute to climatic changes and will add to air pollution that not only affects the health of the Canadian population but have a negative impact on health and economic outcomes nationwide (Israël and Flanagan, 2016).

Electricity Generation Using Coal as Fuel in Canada

Coal is mainly used in Canada for electricity generation and heating and accounts for about 85,3% of the total coal consumption, another 7,2% goes to coke manufacturing, and various industrial uses; other non-energy uses account for 7,6%. Coal-fired generation contributes 9,6% of Canada's total electricity generation in 2016, but it is expected to provide only 0,6% in 2040 (Canada's Energy Future, 2017. Energy Supply and Demand Projections to 2040, 2017).

According to Fig. 4.24, the following can be stated: in 2016, coal was the third energy source for electricity generation and heating in Canada with 9,7% of the total electricity produced in the country in that year (646,3 TWh, including the electricity generated by the industry), after hydro with 58,3% and nuclear power with 14,9%. The expected trend in the electric capacity and generation in percentage and by energy source in Canada in 2040 is shown in Fig. 4.25.

Figure 4.24 Power generation by source. *(Source: Canada's Energy Future (2017). Energy Supply and Demand Projections to 2040 (2017).)*

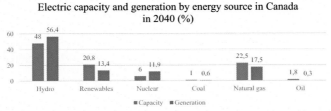

Figure 4.25 Electric capacity and generation in percentage and by energy source in Canada. *(Source: Canada's Energy Future 2017. Energy Supply and Demand Projections to 2040 (2017).)*

Analyzing the data included in Fig. 4.25, it can be stated that renewables and natural gas are the only two energy sources that are expected to increase their participation in the energy mix of the country during the period 2016—40.

Over the years, the progress of environmental policies and actions resulted in reducing the use of coal in coal-fired power plants and, for this reason, the share of electricity generated by coal was also declining. Coal-fired power plants are expected to continue reducing their share of the overall generation mix in the country in the decades to come, either to be replaced with natural gas-fired power plants or other forms of generation such as renewables.

The evolution of the participation of coal in the energy mix of Canada during the period 2010—17 is shown in Fig. 4.26.

According to Fig. 4.26, the participation of coal in the electricity generation and heating in Canada during the period 2010—17 decreased by 17,5% falling from 13,2% in 2010 to 10,9% in 2017.

In 2012, the government of Canada promulgated a new regulation for the use of coal for electricity generation and heating that sets a stringent

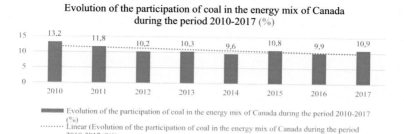

Figure 4.26 Evolution of the participation of coal in the energy mix of Canada during the period 2010—17 (%). *(Source: Natural Resources Canada Electricity Facts, and BP Statistical Review of World Energy (2018) (June 2018).)*

performance standard for CO_2 emissions for new coal-fired power plants and those at their end of life. At the beginning of 2015, the rule had to be fulfilled by a coal-fired power plant with 50-year of operation to be allowed to continue to generate electricity or for new coal-fired power plant to begin generating electricity, if they emit no more than 420 tons of CO_2 per GWh—a rate that is achievable by new baseload natural gas combined cycle technologies available today. The stated objective of the regulations is to change the current structure of the country's energy mix and to shift to lower or nonemitting types of electricity generation, such as high-efficiency natural gas-fired power plants, the use of renewable energy source, or fossil fuel-fired power plants with carbon capture and storage (CCS). In response to the regulation, utilities in Alberta, Saskatchewan, and Nova Scotia will need to decide whether to shut down some coal-fired power plants, replace coal-fired power plants with other technologies, or install CCS technologies (Haffner and Vriesendorp, 2014).

CHAPTER 5

Conclusion

According to the International Energy Outlook 2017 report, world energy use is expected to grow 28% during the period 2015–40. More than a half of this increase is due to the rise in the energy consumption of China and India, among other Asian and Pacific countries, as a result of a robust economic growth that will be registered in those countries during the period analyzed.

On the other hand, according to the International Energy Outlook 2016 with Projection to 2040 report, it is expected that the use of petroleum and other liquid fuels worldwide will grow from 90 million barrels per day in 2012 to 121 million barrels per day in 2040; this means an increase of 34,4% for the whole period analyzed. However, it is important to highlight that most of the growth in liquid fuels consumption is in the transportation and industrial sectors and not in the electricity generation sector. In the transportation sector, liquid fuels will continue to provide most of the energy consumed worldwide at least during the coming years.

The participation of crude oil in the world primary energy consumption decreased by 3% during the period 2005–15 falling from 35,96% in 2005 to 32,94% in 2015. It is expected that this trend will continue without change in the short-term. On the other hand, the participation of natural gas in the world primary energy consumption increased almost 1% during the same period rising from 22,89% in 2005 to 23,85% in 2015. It is expected that the participation of natural gas in the world primary energy consumption will continue this trend during the coming years. Finally, the involvement of coal in the world primary energy consumption decreased by 0,59% falling from 28,61% in 2005 to 28,02% in 2015. It is expected that the participation of coal in the world primary energy consumption will also continue this trend during the coming years and its role in the energy mix of several countries is likely to be lower than today.

The International Energy Agency in its 2016 report also indicated that, in 2015, world crude oil production is projected to reach 94,2 million barrels per day, an increase of 3% with respect to the production level reached in 2014 (2,5 million barrels per day). The expected growth will be of 4,2% or 1,1 million barrels per day in the OECD countries, 3,7% or 1,3

Conventional Energy in North America
ISBN 978-0-12-814889-1
https://doi.org/10.1016/B978-0-12-814889-1.00005-X

million barrels per day in the OPEC member states, and 1,3% or 0,4 million barrels per day in other producing countries.

According to EIA sources, petroleum and other liquid fuels production in countries outside of the OPEC grew by 1,4 million barrels per day in 2015; the main growth occurred in the North America region. Using the EIA database (2018) as reference, the following can be stated: the total US crude oil production decreased by 0,4% during the period 2015–17 falling from 9,408 million barrels per day in 2015 to 9,367 million barrels per day in 2017; this means a decrease of 0,041 million barrels per day. In 2020, the total production of US crude oil is expected to achieve a record figure of more than 10,6 million barrels per day, and is anticipated to continue growing up to 11,3 million barrels per day in 2040; this represents an increase of 6,6% during the period analyzed (see Fig. 5.1).

On the basis of the evolution of the crude oil price during the last years, experts have estimated that the real prices of crude oil in 2016 US dollars will recover from an annual average of less than US$50 per barrel registered in 2017 to more than US$130 per barrel in 2040; this means an increase of US$80 per barrel or 2,6 times compared to the price registered in 2017. In 2018, the crude oil price was forecast to be above US$70, according to the trend in the crude oil market. However, according to OPEC sources, at the end of 2018 the price of crude oil was reported to be, as average, US$65,33 per barrel. It is important to highlight that the US crude oil production was affected by the decline in crude oil prices registered in recent years due to its high production costs.

After 2017, higher crude oil prices, as well as exploration, evaluation, and development programs that expand the operator's knowledge of more

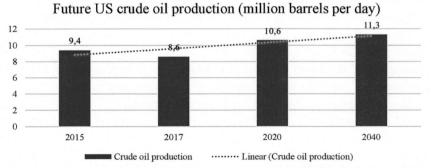

Figure 5.1 Evolution of the projection of the crude oil production per day in the US during the period 2015–40. *(Source: Annual Energy Outlook 2016; US Energy Information Administration, 2018.)*

efficient exploitation of existing crude oil reserves, could result in the identification of additional crude oil resources and the development of new technologies that reduce costs and increase the recovery of existing crude oil reserves in the US.

According to the data included in Fig. 5.1, the US crude oil production is projected to increase by 20,2% during the period 2015—40 rising from 9,4 million barrels per day in 2015 to 11,3 million barrels per day in 2040. However, despite the expected increase in the US crude oil production during the period 2015—40, the crude oil consumption in the country is projected to remain below the 2005 level until 2040.

In the specific case of the US crude oil production in the high seas (which is less sensitive to short-term price movements than onshore production), it is expected to increase up to 2 million barrels per day in 2021, led by new deep-water projects in the Gulf of Mexico,[1] including the Heidelberg and Appomattox fields that started operation in 2016 and 2017, respectively. After 2021, however, crude oil production in the high seas in the US is expected to decrease to approximately 1,6 million barrels per day in 2030 and it is likely to remain at that level until 2040, since the production of newly developed fields will be offset by the anticipated decrease in areas already in operation (Annual Energy Outlook 2016 with Projections to 2040, 2018).

It is forecast that the onshore crude oil production using the recovery of oil improved with CO_2 technique will increase from 0,3 million barrels per day in 2015 to 0,7 million barrels per day in 2040, as oil prices rise and affordable sources of CO_2 are available. It is also expected that crude oil production in Alaska, on both land and sea, will continue to decline, from a total of almost 0,5 million barrels per day in 2015 to less than 0,2 million barrels per day in 2040; that is, a reduction of no less than 40%.

Finally, it is important to stress that despite rising prices, the US crude oil production level is expected to remain the same without significant change, because tight oil development will move into less productive areas and

---

[1] There is increasing evidence that the US sector of the Gulf of Mexico is facing a decline in interest from oil and gas companies due to more lucrative offers in other areas of the country. According to the US Bureau of Ocean Energy Management, there were 14.575 blocks available in the Gulf of Mexico in the latest auction, but only 144 received bids, which is 1% less than those on offer and even four bids lower than the total registered in previous auction (148 bids).

productivity will gradually decrease (Annual Energy Outlook 2017 with Projections to 2050, 2017).

In the case of Canada, it is important to highlight that abundant energy resources have contributed to Canada's position as one of the world's largest energy producers. However, the "long-term success of shale oil and gas greatly depends on the industry's response to the general public's environmental concerns." (Demspter et al., 2016).

Within all available fossil fuels in the North America region, natural gas will be the fastest-growing fossil fuel in the projection period until 2040. Global natural gas consumption is expected to increase by 1,9% per year. Abundant natural gas resources and robust production, including rising supplies of tight gas, shale gas, and coalbed methane, are expected to contribute to the strong competitive position of natural gas in comparison to oil and coal.

Any analysis of the future of natural gas in Canada and the US should consider, among others, the following elements:

- Natural gas is not a renewable energy source and, for this reason, will be exhausted in the future.
- Natural gas is safer and more comfortable to store compared to oil and coal. For this reason, natural gas is, among all fossil fuel sources, the most efficient source of energy for electricity generation and heating, and the only one likely to grow during the coming years. Also, natural gas can be stored in different manners such as in tanks above the ground, in liquid form, or underground, among others.
- Considerable natural gas resources are expected to be found in several areas outside the US. The role of US natural gas will be influenced by the evolution of this market—particularly the growth and efficiency of trade in LNG.
- Natural gas is a reliable energy source, and the supply of natural gas can be ensured, even when a storm is affecting any specific site.
- Natural gas is a highly combustible fuel and, for this reason, mishandling natural gas can lead to a tremendous and destructive explosion.

The most substantial increase in natural gas production during the period 2012—40 is expected to occur in non-OECD Asia (18,7 trillion cubic feet), the Middle East (16,6 trillion cubic feet), and the OECD Americas (15,5 trillion cubic feet). In the US, the production of natural gas is expected to increase by 11,3 trillion cubic feet and will come mainly from shale gas resources (more than half of the US natural gas production), according to the International Energy Outlook 2016 with Projection to 2040 report.

Without a doubt, "shale gas and tight oil are revolutionizing world energy markets. New drilling methods and technologies have suddenly given North America access to vast deposits of oil and natural gas stored in shale and tight rock formations. These resources, largely inaccessible only a decade ago, represent a significant source of economic growth, jobs, and tax revenue" (Dempster et al., 2016).

Also, it is expected that in the US the production of tight gas, shale gas, and coalbed methane will substantially increase during the coming decades. "The application of horizontal drilling and hydraulic fracturing technologies has made it possible to develop the US shale gas resource, contributing to a near doubling of estimates for total US technically recoverable natural gas resources over the past decade" (International Energy Outlook 2016 with Projection to 2040, 2016).

Tight gas, shale gas, and coalbed methane resources in Canada and China are expected to account for about 80% of the total production in 2040. LNG is expected to account for a growing share of world natural gas trade rising from nearly 12 trillion cubic feet in 2012 to 29 trillion cubic feet in 2040; this represents an increase of 241,6%. Most of the rise in liquefaction capacity is foreseeable to occur in Australia and North America, where a multitude of new liquefaction projects are planned or under construction (International Energy Outlook 2016 with Projection to 2040, 2016).

Natural gas production in the OECD Americas is expected to grow by 49% during the period 2012–40. According to the International Energy Outlook 2016 with Projection to 2040 report, the US, which is the largest producer in the OECD Americas and the OECD as a whole, accounts for more than 45,8% of the expected region's total production growth during the period indicated above. The expected increase will be from 24 trillion cubic feet in 2012 to 35 trillion cubic feet in 2040.[2] In 2040, shale gas is expected to account for 55% of the total US natural gas production, tight gas for 20%, and offshore production from the lower 48 states for 8%. The remaining 17% is likely to come from coalbed methane, Alaska, and other associated and non–associated coastal resources in the lower 48 states.

---

[2] US shale gas production is expected to grow 100% from 10 trillion cubic feet in 2012 to 20 trillion cubic feet in 2040, more than offsetting declines in the production of natural gas from other sources.

It is also expected that the use of natural gas for electricity generation and heating in the US and Canada will be higher than the level registered in 2017.

Coal is expected to be the slowest-growing energy source during the projected period in the North America region, and it is likely that it will be surpassed by natural gas by 2030. The use of coal for electricity generation and heating at regional level during the coming years will be lower than the level registered in 2017. In the case of Canada, the government will prohibit the use of coal for electricity generation and heating after 2030. In the US, it is anticipated that the use of coal for electricity generation and heating will decline and its role in the country energy mix in 2040 will be much less than today.

However, it is important to highlight that coal is the most abundant of all available fossil fuels in the world and is used by a variety of sectors, including electricity generation and heating,[3] iron and steel production, cement manufacturing, and as a liquid fuel. Compared with the strong growth in coal consumption in the early 2000s, worldwide coal consumption is projected to remain flat between 2015 and 2040 (about 160 quadrillion Btu). Coal currently fuels 40% of the world's electricity, and it is anticipated to continue to supply a strategic share of power over the next three decades (World Energy Council 2016 Coal, used with the permission of the World Energy Council, www.worldenergy.org, 2016).

The largest coal producing countries are China, the US, India, Indonesia, Australia, and South Africa. India's coal consumption continues to grow by an average 2,6% per year from 2015 to 2040, with the country surpassing the US as the second-largest consumer before 2020 (International Energy Outlook 2017, 2017).

According to Saha (2017), despite the challenges facing the US coal industry by the adoption of new regulations, this type of fossil fuel will remain a viable option for electricity generation and heating in the country for years to come. The US EIA's latest annual report indicates that coal-fired power plants will regain their place at the top of the US electricity generation in 2019 and will hold the position in the 2030s if the CPP is repealed. This forecast is contingent upon expected price trends for other resources used for electricity generation and heating, especially natural gas. The US EIA recently projected a slight increase in domestic coal

---

[3] In the specific case of electricity generation and heating, the current level of coal participation will be almost the same at least during the coming years.

production through 2018 as a result of the expected growth in natural gas prices. However, that augmented production would occur only in the US Western coal mining regions.

It is also important to single out that coal is unlikely to return to its previous role within the US energy mix, even without the application of the CPP regulations. The various energy market forces, such as natural gas and renewable energy projects, will keep a flat demand on the power sector, declining global demand for coal in the future. A survey of utility executives carried out recently in the US found that only 4% think coal use will increase moderately or significantly in their utility's power mix over the coming years, 27% said it would decrease somewhat, and 52% said it would dramatically drop (Saha, 2017).

Over the long-term, the future of the US coal industry will depend on its ability to innovate. Breakthroughs and cost reductions with the introduction of advanced coal technologies, including carbon capture, could reduce the negative perceptions about the negative impact on the environment and human health as a result of the use of coal for electricity generation and heating.

The participation of conventional energy sources in the US energy sector is shown in Fig. 5.2.

According to Fig. 5.2, consumption of oil and other liquids in the US is expected to have the higher percentage in 2040 (32%), followed closely by natural gas (30%) and coal (18%). Although liquid fuels, mostly being oil-based, are expected to remain the largest source of world energy consumption in the coming years, the liquid fuels' share in energy consumption in the US is likely to fall from 37% in 2012 to 32% in 2040; this means a reduction of 5%. In the specific case of natural gas, it is expected to increase

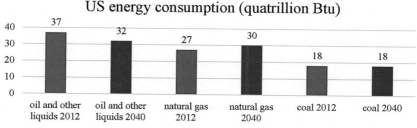

Figure 5.2 Participation of the various energy sources in the US energy sector. *(Source: EIA database 2018.)*

its role from 27% in 2012 to 30% in 2040; this means an increase of 3%. Finally, in the case of coal, it is expected that the participation of this energy source in the US energy mix in 2012 and 2040 be the same.

On the other hand, natural gas production in Canada is expected to grow by 1,2% per year on average over the projection period, rising from 6,1 trillion cubic feet in 2012 to 8,6 trillion cubic feet in 2040. In Canada, like in the US, much of the production growth is expected to come from growing volumes of tight gas and shale gas production (International Energy Outlook 2016 with Projection to 2040, 2016).

The abundance of shale oil and shale gas in Canada "is helping to push down petroleum prices, adding further criticism that higher production is discouraging people from transitioning toward environmentally friendly fuel sources, such as solar, geothermal, and biofuel. On the other hand, low natural gas prices are encouraging more people to convert to clean-burning natural gas" (Dempster et al., 2016).

Finally, in the case of coal (thermal and metallurgical coal[4]) the following can be stated: according to the Canada's Energy Future 2017 report, the total Canadian coal production in 2015 was 61,9 million tons. As the fuel for coal-fired power plants, Canadian thermal coal production is linked to the use of coal in the electricity sector, particularly in Alberta and Saskatchewan. Thermal coal, mostly used for power generation, accounted for 93% of the total coal consumption in 2015 in Canada. However, during the period 2016—40, the demand for thermal coal is expected to decline by 86%, falling from 37,7 million tons in 2016 to 5,4 million tons in 2040.[5] This declining trend is driven primarily by retirements of coal-fired power plants capacity resulting from regulations to phase out old and ineffective coal-fired power plants by 2030. In response to declining domestic demand, production of thermal coal in Canada during the period 2016—40 is also

---

[4] In the case of Canada, "metallurgical coal is primarily used for steel manufacturing domestically and internationally. Much of Canada's metallurgical coal production is exported and future production trends are linked to global coal demand and prices". However, domestic demand for metallurgical coal used in steel manufacturing is expected to be stable over the period 2016—40 at 3 million tons. Global demand for metallurgical coal is expected to grow moderately over the period 2016—40, resulting in steady growth in net exports from Canada. Total metallurgical coal production in Canada is expected to increase by 13,4% from 26,9 million tons in 2016 to 30,5 million tons in 2040 (Canada's Energy Future 2017, 2017).

[5] Most of the thermal coal consumed in Canada in 2040 for electricity generation and heating will be from coal-fired power plants equipped with the CCS technology.

expected to drop 78,2% falling from 36,3 million tons in 2016 to 7,9 million tons in 2040 (Canada's Energy Future 2016. Energy Supply and Demand Projections to 2040, 2017).

It is well-known that electricity generation and heating can be done using different energy sources, which are classified into two big groups. These groups are (1) conventional energy sources, and (2) non- conventional energy sources. The primary conventional energy sources are oil, natural gas, and coal. Non-conventional energy sources include two main groups: (1) renewable, including wind, solar, hydro, biomass, and geothermal energy, among others, and (2) nuclear energy. The North America region is rich in all of the three main conventional energy sources mentioned above.

On the other hand, electricity consumption is an essential component of modern life in all countries. In developed countries, electricity consumption is much higher than the electricity consumption in developing countries. However, in several developing countries, such as China, India, South Korea, South Africa, among others, the electricity consumption is increasing very fast.

Electricity not only provides clean and safe light throughout the day but also refreshes homes on hot summer days and warms homes in winter. In all states, power allows the use of electrical and electronic equipment in which the consumption of this type of energy is essential to ensure proper functioning. It is a reality that without energy people will be deprived of heating, cooling, use of all electronic equipment, and light in their homes and workplaces. They would not have access to television and the internet, among other essential services associated with modern life. With the development of modern societies, the use of these devices and services is expected to grow.

The growth in the world economy means that more energy is required, particularly in the form of electricity. Even though consumption of non-fossil fuels is expected to grow faster than consumption of fossil fuels, this last type of energy source is still expected to account for 78% of the energy use in 2040 (International Energy Outlook 2016 with Projection to 2040, 2016).

References

About Electricity, 2016. Natural Resources Canada.

Aggeliki, K., 2014. Pros and Cons of Natural Gas Use. Bright Hub Engineering.

Alberta's Energy Reserves and Supply/Demand Outlook, 2016. Alberta Energy Regulator; ST98-2016. Retrieved May 20, 2016.

American Geosciences Institute (AGI) Database, 2017.

American Geosciences Institute (AGI) Database, 2018.

Annual Coal Report, November 2017. U.S. Energy Information Administration (EIA).

Annual Energy Outlook 2013, 2013. US Energy Information Administration.

Annual Energy Outlook 2014, 2014. US Energy Information Administration.

Annual Energy Outlook 2015, 2015. US Energy Information Administration.

Annual Energy Outlook 2016. Presentation by Adam Sieminski, Administrator, Independent Statistics & Analysis. www.eia.gov, 2016.

Annual Energy Outlook 2016, 2016. US Energy Information Administration.

Annual Energy Outlook 2016 with Projections to 2040, 2018.

Annual Energy Outlook 2017 With Projections to 2050, 2017. US Energy Information Administration.

Annual Energy Outlook 2018, 2018. US Energy Information Administration.

Austin, S., 2016. Six Reasons Why Oil Prices Reached New 2016 Highs. fossiloil.com/%85/06/08/six-reasons-oil-prices-reached-new-2016-highs.

Bönisch, N. Coal Deposits of the United States of America. TU Bergakademie Freiberg, Germany.

BP Energy Outlook 2016 Edition. Outlook to 2035, 2016.

BP Energy Outlook Focus in North America 2016 edition, 2016. Outlook to 2035.

BP Statistical Review of World Energy 2015, 2016.

BP Statistical Review of World Energy 2017, June 2017.

BP Statistical Review of World Energy 2016, sixty fifth ed., 2016

BP Statistical Review of World Energy 2018, June 2018.

Branson, R., Loewen, E., 2012. Branson Urges Obama to Back Next-Generation Nuclear Technology. http://www.guardian.co.uk/environment/2012/jul/20/richard-branson-obama-nuclear-technology.

Canada's Energy Future 2013 — Energy Supply and Demand Projections to 2035 — an Energy Market Assessment, 2013. National Energy Board.

Canada's Energy Future 2016. Energy Supply and Demand Projections to 2040, National Energy Board, 2016.

Canada's Energy Future 2016. Energy Supply and Demand Projections to 2040, 2017.

Canada's Energy Future 2017. Energy Supply and Demand Projections to 2040, 2017. National Energy Board, An Energy Market Assessment.

Canada Energy Policy Laws and Regulations Handbook. Strategic Information and Basic Laws, vol. 1, March 3, 2015. World Business and Investment Library, Lulu, Inc. IBP.

Canada: EIA Country Overview, 2014. Energy Information Administration.

Canada International Data and Analysis, 2015. US Energy Information Administration (EIA).

Canada's Natural Gas, 2018. Canada's Oil and Natural Gas Producers.

Canada Oil Market Overview, 2016. Energy Information Administration.

Clemente, J., 2015. US Oil Production Forecasts Continue to Increase; Forbes Energy.

Coal Explained. Coal Imports and Exports, 2018. EIA.

Coal and Environment, 2018. Energy Information Administration.

Coal Facts, 2017. Natural Resources Canada.

Coal Facts, 2018. Natural Resources Canada.

Coal-Fired Power Plants and the Menace of Mercury Emissions, August 2001. A Greenpeace Southeast Asia Report.

Coal Power and Air Pollution, 2008. Union of Concerned Scientists.

Coal and Air Pollution, 2017. Union of Concerned Scientists.

Council of Canadian Academies, 2014. Environmental Impacts of Shale Gas Extraction in Canada (2014); the Expert Panel on Harnessing Science and Technology to Understand the Environmental Impacts of Shale Gas Extraction. Council of Canadian Academies.

Cost and Performance Baseline for Fossil Energy Plants. Bituminous Coal and Natural Gas to Electricity, vol. 1, Revision 2, November, 2010. National Energy Technology Laboratory (NETL), DOE/NETL-2010/1397, United States Department of Energy.

Creating Cleaner Air in Ontario: Province Has Eliminated Coal-Fired Generation, 2014. Ontario Ministry of Energy. http://news.ontario.ca/mei/en/2014/04/creating-cleaner-air-in-ontario-1.html.

Crude Oil Facts, 2018. Natural Resources Canada.

Crude Oil Annual Export Summary; National Energy Board, 2018.

Crude Oil Forecast, Markets, and Transportation, 2017. Canada Association of Petroleum Producers (CAPP).

Crude Oil & Canada, 2016. Dean Monterey, Global Incident Command Solutions, Inc.

Crude Oil Production (2014); EIA; http://www.eia.gov/dnav/pet/pet_crd_crdpdn_adc_mbbl_m.htm.

Crude Oil, CO_2 and Products Pipelines (2007); REXTAG Strategies Corp and the Alaska Department of Natural Resources; EIA; 2007.

Choudhury, N., November 06, 2017. Oil & Gas Industry Outlook: Nasdaq. http://www.nasdaq.com/article/oil-gas-industry-outlook-november-2017-cm872707.

Cycling of Mercury to the Global Environment (2002) UNEP Global Mercury Assessment; 2002.

Dempster, J., Goupil, S., Couture, I., January 2016. Canada's Next Energy Frontier: Shale Oil and Gas. Natural Resources Canada.

Department of the Interior and US Geological Survey database.

Dhillon, K., 2014. Why Are the US Oil Imports Falling? Time.

Durden, T., December 29, 2016. Can the Canadian Oil Industry Recover in 2017?

Effects of Acid Rain: Human Health, 2004. US EPA (2003). www.epa.gov/airmarkets/acidrain/effects/health.html.

EIA, 2007. Crude Oil, CO_2 and Products Pipelines and Refineries.

EIA Database, 2016.

EIA Database, 2017.

EIA Database, 2018.

Electricity Facts 2018, 2018. Natural Resources Canada.

Electricity Data Browser, 2018. EIA.

Electricity Trade Summary of the National Energy Board of Canada, 2018.

Electricity Power Annual 2015, 2016. US Energy Information Administration (EIA).

Electricity Power Annual 2016, 2018. US Energy Information Administration (EIA).

EIA Electric Power Monthly, 2018. EIA.

Energy Fact Book 2016—2017, 2016. Natural Resources Canada.

Energy Markets Fact Book-2014/2015, 2014.

Energy Policy of Canada, 2018. Wikipedia.

Energy Policies of IEA Country Review, 2015. International Energy Agency.

Energy in the United States, 2018. Wikipedia.

Erbach, G., 2014. Unconventional Gas and Oil in North America: The Impact of Shale Gas and Tight Oil on the US and Canadian Economies and on Global Energy Flows. European Parliamentary Research Service.

Estimated Natural Gas Pipelines Mileage in the Lower 48 States, 2008. Energy Information Administration.

Exploration and Production of Shale and Tight Resources, 2016. Natural Resources Canada.

Gas in the United States of America, 2018. World Energy Council.

Gerhardt, T., 2012. Record Number of Coal Power Plants Retire; E-Magazine. Archived from the original on 1 November 2012.

Global Trends in Oil & Gas Markets to 2025, 2013. Lukoil.

Global Energy Statistical Yearbook 2018, 2018. Enerdata.

Gupta, Deepali; Unconventional Petroleum; https://www.scribd.com/presentation/360377790/Unconventional.

Haffner, J., Vriesendorp, W., April 2014. Vision 2050. The Future of Canada's Electricity System. Canadian Electricity Association.

History of Electricity, August 29, 2014. Institute for Energy Research.

Hood, M., June 13, 2016. Cheap Gas, Coal Won't Hobble Renewables: Energy Report; Paris (AFP).

Houser, T., Bordoff, J., Marsters, P., 2017. Can Coal Make a Comeback? Center on Global Energy Policy, School of International and Public Affairs, Columbia University. Retrieved May 15, 2017. energypolicy.columbia.edu.

How Much Coal Does the US Exports and Import? 2017. American Geosciences Institute.

How Much Oil Is Consumed in the United States? 2018. Energy Information Administration.

IEA, 2017. Coal Information 2017. OECD Publishing, Paris. https://doi.org/10.1787/coal-2017-en.

Indexmundi Database.

Ingraffea, A., Wells, M.T., Santoro, R.L., Shonkoff, S.B.C., 2014. Assessment and risk analysis of casing and cement impairment in oil and gas wells in Pennsylvania, 2000—2012. Proceedings of the US National Academic of Science 111 (30).

Installed Plants, Annual Generating Capacity by Type of Electricity-Generation; Statistics Canada, 2018.

International Energy Agency, 2016. IEA.

International Energy Outlook 2016, May 2016. Energy Information Administration (EIA).

International Energy Outlook 2017, September 2017. Energy Information Administration (EIA).

International Energy Outlook 2016 with Projection to 2040, 2016. US Energy Information Administration, Office of Energy Analysis, US Department of Energy, IEO 2016.

International Trade; www.intracen.org/itc/market-info-tools/trade-statistics; 2018.

Institute for Energy Research Database, 2018.

Israël, B., Flanagan, E., November 2016. Out with the Coal, in with the New; National Benefits of an Accelerated Phase-Out of Coal-Fired Power. Pembina Institute.

Keating, M., 2001. Cradle to the Grave: The Environmental Impacts from Coal. www.catf.us/publications/reports/Cradle_to_Grave.pdf.

Kolstad, C.D., 2017. What Is Killing the US Coal Industry? Institute for Economic Policy Research (SIEPR), Stanford University.

Kristopher, G., 2017. What US Crude Oil Production's 26 — Month High Could Mean. Market Realist.

Magill, J., 2015. Canadian Natural Gas E&P Outlook Weak: Analyst. Natural Gas.

Manzagol, N., Hodge, T., 2015. US-Canada Electricity Trade Increases. Toda Energy, EIA.

McGuire, P.A., 2001. Coal Gets Cleaner and Better Connected. Businessweek Online.

Medium-Term Coal Market, 2015. International Energy Agency.

Medium-Term Oil Market, 2015. International Energy Agency.

Mercator Research Institute on Global Commons and Climate Change, 2015. Renaissance of Coal Isn't Stopping at China.

Moving Crude Oil by Rail, 2014. Association of American Railroads.

Mufson, S., 2017. Trump Seeks to Revive Dakota Access. Keystone XL Oil Pipelines, Washington Post, Washington, DC. Retrieved January 24, 2017.

Mutemi, A., 2013. MUI Coal Mines: A Blessing or a Curse? Socio-Economic and Environmental Intricacies. University of Nairobi, School of Law, 2013

MxKinnon, H., Muttitt, G., Stockman, L., 2015. Figure 1: Main Pipeline and Proposed Pipeline Routes Leading Out of the Alberta Tar Sands, Lockdown: The End of Growth in the Tar Sands, p. 11. City: Oil Change International, 2015. http://priceofoil.org/.

National Resources Canada, Electricity Facts, 2015.

National Resources Canada, 2017.

National Energy Board; Commodity Statistics, 2018

Natural Gas Imports and Exports, 2016. US Energy Information Administration (EIA).

Natural Gas Facts, 2018. Natural Resources Canada.

Natural Gas Information: Overview 2017, 2017. International Energy Agency, IEA.

Natural Gas Pipelines, 2017. US Energy Information Administration (EIA), Natural gas explained.

Natural Gas Pipeline System in the United States, 2018. Wikipedia.

Norris, F., 2014. US Oil Production Keeps Rising beyond the Forecasts. The New York Times.

Nyquist, S., 2018. The US Growth Opportunity in Shale Oil and Gas, Interview Transcript. https://www.mckinsey.com/industries/metals-and-mining/our-insights.

Oil, 2017. Institute for Energy Research (IER).

OPEC Database, 2016.

OPEC Database, 2017.

OPEC Database, 2018.

Ovale, P., December 11, 2014. Her ser du hvorfor oljeprisen faller. English Teknisk Ukeblad.

Petroleum in the United States Wikipedia, 2018

Petroleum Industry in Canada, 2018. Wikipedia.

Petroleum Resources Management System, 2007. Society of Petroleum Engineers (SPE), American Association of Petroleum Geologists (AAPG), World Petroleum Council (WPC), and Society of Petroleum Evaluation Engineers (SPEE).

Pipelines Across Canada, 2016. Natural Resources Canada.

Potenza, A., 2017. The EPA Will Reverse a Critical Clean Energy Policy so Polluters Can Burn More Coal. Bye-bye, Clean Power Plan.

Quigley, J., February 04, 2016. Coal in Canada: A By-The-Numbers Look at the Industry. CBC News. Online at. http://www.cbc.ca/news/business/canadian-coal-by-the-numbers-1.3408568.

Reduction of Carbon Dioxide Emissions from Coal-Fired Generation of Electricity Regulations, 2012. Government of Canada. http://www.gazette.gc.ca/rp-pr/p2/2012/2012-09-12/html/sor-dors167-eng.html.

Reinsalu, E., Aarna, I., 2015. About technical terms of oil shale and shale oil. Oil Shale 32 (4), 291−292. https://doi.org/10.3176/oil.2015.4.01. ISSN 0208-189X, Estonian Academy Publishers.

Saha, D., September 11, 2017. Trends and Market Forces Shaping the Future of U.S: Coal Industry. The Council of State Governments. Online at. http://knowledgecenter.csg.org/kc/content/trends-and-market-forces-shaping-future-us-coal-industry.

Shale Oil: The Next Energy Revolution, 2013. Silverback Exp. http://www.silverbackexp.com/sites/default/files/resources_pwcshaleoil.pdf.

Shale Gas Production Drives World Natural Gas Production Growth, 2016. Today in Energy, EIA.

Short-Term Energy Outlook, 2017. US Energy Information Administration.

Short-Term Energy Outlook, May 2018. US Energy Information Administration.

Sieminsky, A., 2015. In: Oil and Natural Supply and Demand Trends in North America and Beyond; Energy Metro Desk Conference; New Risk in Energy II. EIA, Houston, US.

Speight, J.G., 2012. The Chemistry and Technology of Coal, third ed. CRC Press, ISBN 9781439836460, p. 13.

Springfield, C., 2016. The Decline in the US Coal Industry. International Banker.

Statista Database, 2017.

Statista Database, 2018.

Statistics Canada, 2015.

Statistics Canada, 2018.

Statistics Canada 2018, 2018. Natural Resources Canada.

Statistical Handbook for Canada's Upstream Petroleum Industry, September 2015. Canada Association of Petroleum Producers. http://www.capp.ca/publications-and-statistics/statistics/statistical-handbook.

Swier, R., 2018. U.S. Energy Facts in Energy, Environment, Featured, Policy, Politics, Regulation, Social Issues.

Schneider, C.G., 2000. Death, Disease and Dirty Power: Mortality and Health Damage Due to Air Pollution from Power Plants; Clean Air Task Force. www.catf.us/publications/reports/Dirty_Air_Dirty_Power.pdf.

Tarr, J.A., 2014. Toxic Legacy: The Environmental Impact of the Manufactures Gas Industry in the United States, vol. 55(1). Carnegie Mellon University, Department of History, Technology and Culture, pp. 107−147.

Taylor, J., 2016. Trump's Energy Policy: 10 Big Changes.

Technically Recoverable Shale Oil and Shale Gas Resources: Canada, 2015. EIA, Independent Statistics and Analysis.

Tertzakian, P., 2012. Canada Again a Focus of a New Great Scramble for Oil. The Globe and Mail.

The Environmental Impacts of Coal, May 2005. Greenpeace Briefing; Climate, New Zealand.

The Future of Natural Gas, 2010. An Interdisciplinary MIT Study.

The Socioeconomic Impacts of Advanced Technology Coal-Fueled Power Stations, 2015. Coal Industry Advisory Board to the IEA, Paris.

The United States Is a Net Energy Importer from Canada, 2018. Geniux World Latest News.

Tierney, S.F., 2016. The U.S. Coal Industry: Challenging Transitions in the 21st Century. Analysis Group Inc.

Tight Oil Developments in the Western Canada Sedimentary Basin, 2011. Energy Briefing Note, National Energy Board.

Today in Energy, 2016. US Energy Information Administration (EIA).

Today in Energy, 2017. Canada is the United States' Largest Partner for Energy Trade, EIA.

Today in Energy, 2018. US Energy Information Administration (EIA).

United States Department of State Bureau of Oceans and International Environmental and Scientific Affairs, 2013. Draft Supplemental Environmental Impact Statement for the Keystone XL Project Applicant for Presidential Permit: TransCanada Keystone Pipeline. United States Department of State. Retrieved March 17, 2013.

Uranium 2014, 2015. Resources, Production and Demand. OECD, NEA.

US Crude Oil and Natural Gas Proved Reserves Year-End 2016, 2018. EIA.

US Crude Oil Exports Are Increasing and Reaching More Destinations, 2016. EIA.

US Department of Labor, Mine Safety and Health Administration — MSHA, 2017. http://arlweb.msha.gov/stats/centurystats/coalstats.asp.

US Department of Labor — Bureau of Labor Statistics, 2018. http://www.bls.gov/iif/oshwc/osh/os/osar0012.htm.

US Natural Gas Imports and Exports for 2016, 2017. US Energy Information Administration (EIA).

US Rail Crude Oil Traffic, 2017. Association of American Railroads.

Venkata Lalitha, D., Ujwala, P., Bhuvaneswari, A., 2014. A Comparative Study Paper: Power Generation -Conventional and Renewable Energy Sources. College: Dr.L.Bullayya College of Engineering (for women), Visakhapatnam.

Walsh, C., 2017. Canada's Oil Reserves 2nd Only Toto Saudi Arabia; Dow Jones Newswires. Petroleumworld.com.

What Is Natural Gas? 2018. Canada Natural Gas.

Western Canadian Oil and Gas, Exploration and Production Industry, 2014. Canada: EIA Country Overview. Energy Information Administration.

World Energy Resources 2016, 2016. World Energy Council.

World Energy Resources 2016 Coal, 2016. World Energy Council.

Executive Summary, used by permission of the World Energy Council World Energy Resources 2016, 2016. www.worldenergy.org.

World Energy Outlook 2001, 2001. International Energy Agency.

World Energy Outlook 2012, 2012. International Energy Agency.

World Energy Outlook 2014, 2014. International Energy Agency.

World Energy Outlook 2015, 2015. International Energy Agency.

World Energy Outlook 2016, 2016. International Energy Agency.

World Steel Association, Steel Statistical Yearbook 2015; Brussels, Belgium, 2015, pp. 91—92. https://www.worldsteel.org/statistics/statisticsarchive/yearbook-archive.html.

World Nuclear News, 2017.

Xu, C., Donahue, T., Slocum, M.T., Bell, L., December 2015. Reserves grow modestly as crude oil production climbs. Oil & Gas Journal 113 (12).

Index